GCSE
MATHS
PRACTICE 2

Rosalind Hollis

CASSELL

Cassell Publishers Limited
Artillery House
Artillery Row
London SW1P 1RT

First published 1989

British Library Cataloguing in Publication Data

Hollis, R. G. (Rosalind G.)
   GCSE maths practice 2.
   1. Mathematics
   I. Title
   510

ISBN 0-304-31750-0
         0-304-31751-9 (with answers)

Typeset in Futura and Times Roman by Fakenham Photosetting Limited,
Fakenham, Norfolk
Printed and bound in Great Britain by Billing & Sons Ltd., Worcester

Designed by Vaughan Allen

contents

preface

This series is written for the prospective lower and middle GCSE ability range (grades D to G). All the topics required for these levels by the various examination boards are included.

The items from List 1 of the National Criteria (1985) are covered thoroughly. It is based on the belief that both mechanical practice and problem solving are important. All the items from List 2 are covered using simple examples and applications appropriate to pupils who may be entered at the intermediate level.

The basic processes are covered thoroughly in graded exercises which allow pupils to gain confidence and demonstrate positive achievement. The worked examples at the top of the pages will serve as a reminder. They do not presuppose any teaching method. All the work on each part of the syllabus is grouped together as far as seems appropriate at this level. Teachers will wish to make selections according to their pupils' needs. Each page is self contained to allow as much flexibility as possible.

Some exercises are written to encourage the use of approximations and to develop a feel for number, length, weight, etc. Intelligent use of a calculator is encouraged. Practical and everyday situations are widely used. Familiarity with both metric units and the imperial measures in common use is assumed. Clear and direct questions are used to encourage pupils to start exploring each problem. Some of the situations lend themselves to class discussion.

There is a collection of assorted questions. Some are the type which occur in examinations and others may be used as the starting point for investigations.

The papers of mixed practice have been constructed so that there are parallel papers. Thus, if children have done a paper, and then worked on the parts they found difficult, they can then do the next paper with confidence that their score will be much improved because it covers exactly the same material.

R. G. Hollis, October 1988

1 number

DOMINOES

A

1 Draw the four missing dominoes.
2 How many 6's in the set?
3 How many 5's in the set?
4 How many 4's, 3's, 2's, 1's, 0's?

B

How many 6's have been played? How many 6's are left?

C

How many 5's have been played? How many 5's are left?

D

How many 6's have been played?
How many 6's are left?

Where should they be placed so the game is finished?

E Karen and Lucy decided to play their own
version. They removed all the pieces with a 5 or 6.
How many pieces were left?

This is how the first 4 pieces were played:
How many 2's are left?

It is Karen's turn and she decides to play
one of these pieces:
Which is it best for her to play?

1

SQUARE NUMBERS

A

1 Copy and complete this table:

Number of stars in a row	1	2	3	4	5	6
Number of stars in a column	1	2				
Number of stars in the block	1	4				

2 List the square numbers below 100.

3 Look at this way of writing square numbers:

$1 = 1^2 = 1$
$4 = 2^2 = 1 + 3$
$9 = 3^2 = 1 + 3 + 5$
$16 = 4^2 = 1 + 3 + 5 + 7$
$25 = 5^2 = 1 + 3 + 5 + 7 + \ldots$

How many of the odd numbers are added together to make 25?
How many of the odd numbers are added together to make 49?

B These are the triangular numbers:

1 Call the first triangular number T_0, the next T_1 and so on. Copy and complete this table, using the pattern to find the next two triangular numbers:

Number of stars in bottom row	1	2	3	4	5	6	7	8
Number of stars in triangle	1	3	6					
	T_0	T_1	T_2	T_3				

2 Copy and complete this:

$1 \qquad\qquad = 1$
$1 + 2 \qquad\quad =$
$1 + 2 + 3 \qquad =$
$1 + 2 + 3 + 4 \quad =$
$1 + 2 + 3 + 4 + 5 \ =$

Write two more lines to show how the triangular numbers T_5 and T_6 are made.

3 Calculate $T_0 + T_1$
$\qquad\quad T_1 + T_2$
$\qquad\quad T_2 + T_3 \qquad$ and so on until another pattern appears.

2

NUMBERS FROM SHAPES

	Number
L_1	2
L_2	3
L_3	4

This shows the linear numbers
L_0 L_1 L_2 L_3 L_4 L_5 L_6 L_7

	Number
T_1	3
T_2	6
T_3	

	Number
S_1	4
S_2	
S_3	

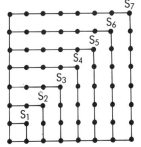

All the numbers on this page are found by counting the dots.
Always try to find a pattern to save the amount of counting!

1 Make a list or table of the linear numbers up to L_7.
2 Make a list or table of the triangular numbers up to T_7.
3 Make a list or table of the square numbers up to S_7.
4 Make a list or table of the pentagonal numbers up to P_7.
5 Make a list or table of the hexagonal numbers up to H_7.

6 Calculate
$L_1 + P_1$ and $T_1 + S_1$
$L_2 + P_2$ and $T_2 + S_2$
$L_3 + P_3$ and $T_3 + S_3$
$L_4 + P_4$ and $T_4 + S_4$
$L_5 + P_5$ and $T_5 + S_5$ etc.

7 Calculate
$T_1 + H_1$
$T_2 + H_2$
$T_3 + H_3$ etc.
Which other pair give the same totals as $T_{17} + H_{17}$?

	Number
P_1	5
P_2	12
P_3	22

	Number
H_1	6
H_2	15
H_3	28

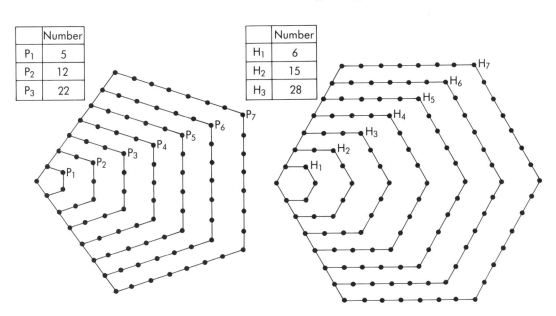

PASCAL'S TRIANGLE

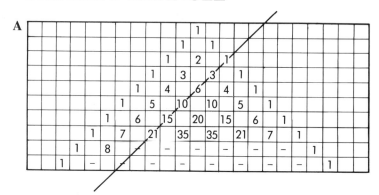

A

1 Copy this pattern of numbers on to squared paper.

2 Fill in the missing numbers.

3 The line goes through the triangular numbers. Make a list of them.
Draw a line through the other line of triangular numbers.

4 Calculate
1 + 3 + 6
1 + 3 + 6 + 10
1 + 3 + 6 + 10 + 15
Where do these numbers appear in Pascal's Triangle?

5 Calculate
1 + 2 + 3 + 4
1 + 5 + 15 + 35

Where do these numbers and their totals appear in the Triangle?
Write some more sums like these.

6

1	=1
1 + 1	=
1 + 2 + 1	=
1 + 3 + 3 + 1	=
1 + 4 + 6 + 4 + 1	=
1 + 5 + ...	=

Use the pattern of the totals to find the total of the tenth line. Find the totals of some more lines.

MOVES ON A GRID

B

The moves allowed are: down and right (coded D and R).
The route RDR reaches A. Route DRR also reaches A.
What is the third route to A?
Copy the grid and label the point A 3 because there are just 3 routes from the start to A.

The route RRR reaches B. There is no other route to B so it is labelled 1.
Find the number of routes to each point and write the numbers on the grid.
If the grid is extended and more points are labelled a pattern should emerge.

4

PATTERN ON SQUARED PAPER

1 In the centre of a large sheet of squared
paper Jane filled in three black squares:

On the side of each black square she then
drew a red square.
How many red squares did she draw?

On the side of each red square she then
drew a green square.
How many green squares did she draw?

Jane is going to put a yellow square against
each green one. How many yellow squares
will there be?

Put all the results in this table:

Colour	1st Black	2nd Red	3rd Green	4th Yellow	5th	6th	7th		
Number	3	8							

Continue filling in the table using the pattern of numbers.
Use the pattern to find the number of squares to be coloured in the twentieth colour.

2 Mary decided to start with two black squares.
She then followed Jane's rules and drew a red square against each black one, and so on.
Was there a pattern in her results?

3 Jill used the same rules and tried starting with just one black square.
How did her pattern compare with those of Jane and Mary?

USING ONES AND TWOS

1 Copy these patterns and complete them:

$1=1$

$2=1+1$
$2=2$
▶ 2 ways

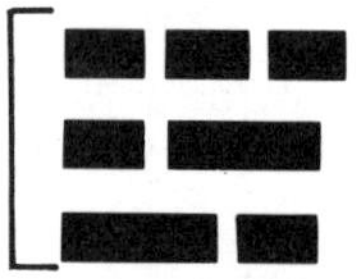

$3=1+1+1$
$3=1+2$
$3=2+1$
▶ 3 ways

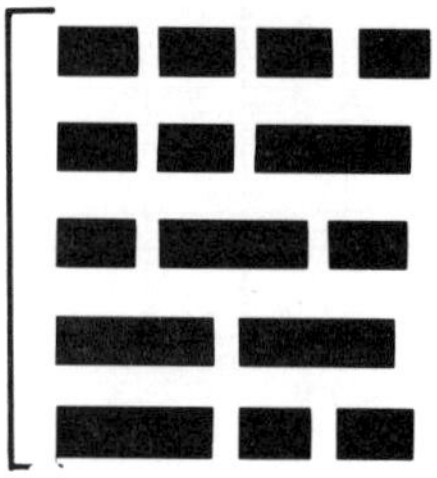

$4=1+1+1+1$
$4=1+1+2$
$4=1+2+1$
$4=2+2$
$4=2+1+1$
▶ 5 ways

$5=1+1+1+1+1$
$5=2+1+1+1$
$5=1+2+1+1$
$5=$
$5=$
$5=$
$5=$
$5=$
▶ ... ways

$6=1+1+1+1+1+1$
$6=$
$6=$
$6=$
$6=$
▶ 13 ways

2 There are 21 ways of making 7 from 1's and 2's. Can you find them all?

3 Look at the pattern of the number of ways:
1, 2, 3, 5, ..., 13, 21.

$1+2=...,$ $2+3=...,$ $3+5=...,$ $5+...=13.$

Write four more numbers in this pattern.

6

PLAYING WITH NUMBERS

1 Tom arranged his model soldiers in groups of 4 and had 2 left. He has less than 50 altogether. Make a list of the possible numbers of soldiers in his collection.
He then rearranged them so there were 5 in each group and 1 was left alone. Again, make a list of the possible numbers he may have had.
Each time he arranged the soldiers there were more than 3 groups. How many soldiers did he have?

2 Mr Smith tells his class that he is not yet 60 and his age this year is a multiple of 7. Make a list of his possible ages.
Then he says that next year his age will be a multiple of 5. Make a list of those possible ages. Rewrite this second list to show how old he might be this year.
How old is Mr Smith?

3 Ann and Bill shake hands. A———B one handshake.

Carol arrives and shakes hands with Ann and Bill: 3 people, 3 handshakes.

If David had arrived at the same time and everyone had shaken hands with everyone else, how many handshakes would there have been?

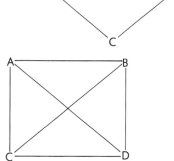

Draw a diagram to show the handshakes between 5 people.
Display the numbers in a table:

Number of people	1	2	3	4	5	6
Number of handshakes	0	1				

4

In a game 3 players try to make words using all or some of these letters.
The first player makes

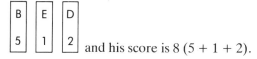

and his score is 8 (5 + 1 + 2).

If he had made BEND, what would the score have been?
Calculate the score for BONED.
What is the lowest possible score for a three letter word?
What is the score for BROWN?
What is the highest possible score for a seven letter word?
The third player scores 8 for a word using O, R and N. What is the other letter?

EXAMPLE

$56 \times 300 = 5600 \times 3$ $748 \div 40 = 74.8 \div 4$ $14.52 \times 40 = 145.2 \times 4$ $2250 \div 300 = 22.5 \div 3$
         $= 16800$ $= 18.7$ $= 4180.8$ $= 7.5$

A Calculate:

1. 36×200
2. 321×20
3. 42×500
4. 36×40
5. 972×200
6. 635×600
7. 502×20
8. 23×2000
9. 41×80
10. 750×300
11. $604 \div 50$
12. $366 \div 600$
13. $1260 \div 4000$
14. $378 \div 90$
15. $3080 \div 70$

B Calculate:

1. 32.1×60
2. 42.3×300
3. 56.4×50
4. 231.5×50
5. 47.1×700
6. 5.6×2000
7. 90.3×300
8. 0.9×400
9. 0.05×60
10. 0.35×40
11. $630 \div 300$
12. $5.64 \div 40$
13. $0.084 \div 70$
14. $0.656 \div 200$
15. $700 \div 2000$

Find an approximate answer and the exact value.
Show the numbers chosen for the approximation.

C
1. 15.2×201
2. 27.9×49
3. 149×68
4. 702×7.82
5. 35.6×302

6. $258.4 \div 304$
7. $2923 \div 39.5$
8. $36411 \div 68.7$
9. $27339 \div 701$
10. $59600 \div 594$

D
1. What is the next whole number after 7809?
2. Write the next number in the sequence
   300,   30,   3,   0.3, . . .
3. Write the next number in the sequence
   750,   75,   7.5, . . .

4. Write the next number in the sequence
   0.06,   6,   600, . . .
5. Write the next number in the sequence
   100,   400,   900,   1600,   2500, . . .

E Use $<$, $>$ or $=$ to make true statements. Do not use a calculator.

1. 28.9×3 90
2. 30.9×6 180
3. 70.09×8 560
4. 1.078×9 9
5. 19.4×7 150

6. 34.5×2 69
7. 0.75×40 30
8. 1.79×300 600
9. 0.8×51 40
10. 3.04×80 25

11. 5.61×20 113
12. 0.81×7 5.6
13. 9.237×20 180
14. 0.07×49 3.5
15. 8.95×300 2700

EXAMPLE

The multiples of 5 are 5, 10, 15, . . .
2, 3, 5, 7, 11, 13, . . . are prime numbers

The square of 6 = 36
The cube of 4 = 64

A Write down the numbers in this part of the number line which are:

1 the multiples of 4.
2 the multiples of 6.
3 the odd multiples of 5.
4 the even multiples of 7.
5 the odd multiples of 3.

6 prime numbers.
7 square numbers.
8 multiples of 10.
9 in the sequence 3, 8, 13, 18.
10 in the sequence 32, 29, 26, 23.

B

	13		21		25		27		29
30		35		40		45		48	
	49		50		54		60		64

From this set of numbers write down:

1 all the prime numbers.
2 the odd multiples of 7.
3 the square numbers.
4 the multiples of 6.
5 numbers in the sequence 10, 20, 30, . . .

6 the odd multiples of 9.
7 the largest multiple of 8.
8 the smallest odd number.
9 the number 7 greater than 22.
10 the number 9 less than 63.

C Only one straight line can be drawn through the 2 points in A.

1 How many straight lines can be drawn through the 3 points in B?
2 How many straight lines can be drawn through the 4 points in C?
3 Be sure to find all the possible straight lines which can be drawn through the points.

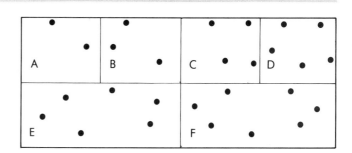

Copy the table and enter the rest of the results:

Number of points	2	3	4	5	6	7				
Number of lines	1	3								

4 What is the name given to the bottom row of numbers?

5 Use the pattern to fill more spaces in the table.

EXAMPLE

The product of 3 and 5 is 15; $3 \times 5 = 15$

The factors of 15 are 3, 5 and 15 because
$15 \div 3$ and $15 \div 5$ are whole numbers.

The square root of 16 is 4, because $4^2 = 16$

The factors of 20 are 2, 4, 5, 10 and 20

A
1. Write 12 as a product of two numbers.
2. What are the prime factors of 30?
3. What is the square root of 9?
4. What is the square of 9?
5. What is the square of 6?
6. What is the largest factor of 60?
7. What is the cube root of 8?
8. What is the square root of 49?
9. What is the cube of 3?
10. What is the next prime after 29?

B

65	3	15
40	17	60
75	20	

From these numbers write down:

1. the prime numbers.
2. a factor of 60.
3. any two which have a total of more than 100.
4. a pair whose total is exactly 100.
5. a pair which differ by 3.
6. three whose total is exactly 100.
7. three numbers to make this a true statement: $A \times B = C$.
8. three numbers to make this a true statement: $P + Q < R$.
9. two numbers to make this a true statement: $X - Y < 15$.
10. a number to make this a true statement: $3 \times 17 < E$.

C
1. $A \times 5 = 304$. A is a whole number. Without doing a calculation how do you know it is the wrong answer?
2. $B \times 300 = 9060$. B is a whole number. Without doing a calculation how do you know this is the wrong answer?
3. C and D are both odd numbers. How can you be sure that their product is not 460?
4. What can be said about the product of two even numbers?
5. What can be said about the product of two even numbers and an odd number?
6. Which of these is the best estimate for the square root of 65:   32,   7.9,   8?
7. Use $<$ or $>$ to make a true statement using the square of 3 and the cube of 2.
8. The sum of two whole numbers is 573. One of them is odd. What is known about the other number?
9. Add twelve thousand and sixty to one thousand and forty.
10. What is fifteen thousand and ten to the nearest hundred?

10

EXAMPLE

63.76 is: 64 to the nearest whole number
       63.8 to one decimal place
       60 to one significant figure
       63.8 to three significant figures
       60 to the nearest ten

Standard form:
$714.6 = 7.146 \times 10^2$
$59.7 = 5.97 \times 10$
$0.826 = 8.26 \times 10^{-1}$

A Make these approximations:

1 4.8 to the nearest whole number
2 48.29 to the nearest whole number
3 10.46 to the nearest whole number
4 0.85 to the nearest whole number
5 0.09 to the nearest whole number
6 11.67 to one place of decimals
7 0.309 to one place of decimals
8 14.351 to one place of decimals
9 1.245 to two places of decimals
10 0.038 to two places of decimals
11 250.6 to the nearest ten
12 4468 to the nearest ten
13 5367.9 to the nearest ten
14 4702 to the nearest ten
15 5099 to the nearest thousand
16 seven hundred and eight to the nearest ten
17 fifteen thousand and ninety-nine to the nearest ten
18 five thousand seven hundred to the nearest thousand
19 134 561 to the nearest ten
20 799 899 to the nearest hundred

B Approximate:

1 53.6 to 2 significant figures
2 536 to 2 significant figures
3 785.1 to 2 significant figures
4 5575 to 2 significant figures
5 0.967 to 2 significant figures
6 0.027 to 1 significant figure
7 5.09 to 1 significant figure
8 756 to 1 significant figure
9 472.9 to 3 significant figures
10 4008 to 3 significant figures

Write in standard form:

11 160
12 460.6
13 500
14 1200
15 35.2
16 0.8
17 0.076
18 0.2
19 0.932
20 0.005

C Calculate:

1 4.56×3
2 3.45×6
3 0.37×2
4 0.24×6
5 1.03×5
6 12.03×5
7 206.8×3
8 931.5×2
9 0.0063×7
10 1.008×3

11 $3.75 \div 3$
12 $11.6 \div 4$
13 $1.15 \div 5$
14 $0.18 \div 3$
15 $0.45 \div 9$
16 $0.0036 \div 6$
17 $53.5 \div 5$
18 $0.272 \div 4$
19 $10.4 \div 8$
20 $24.36 \div 6$

1 Write in order of size, smallest first:
0.33, 0.375, 0.044, 0.3.

2 Put the correct numbers in place of the stars:

$$
\begin{array}{r}
5 \\
+\ 725 \\
\hline
10*1 \\
\hline
\end{array}
\qquad
\begin{array}{r}
7*61 \\
-28*4 \\
\hline
73 \\
\hline
\end{array}
$$

3 Use the signs <, > or = to make a true statement connecting these numbers:
2.56, 1.99, 19.9 and 5.1.

4 Write down two numbers which add up to 5.3.

5 The difference between two numbers is 1.5.
What could the two numbers be?

6 Write down two numbers which multiply to give 2.4.

7 Two whole numbers are divided and the answer is 0.6.
What could the numbers be?

8 Find two numbers which make 7.5 when they are multiplied.

9 Tom puts a number into a calculator and then performs the following operations:

(a) If Tom entered the number 4, what is the answer obtained after doing the above operations?
(b) What number would he have to enter to obtain the number 23?

10 A calculator was used to divide two numbers. What could the numbers have been if the answer that appeared in the display was
(a) 0.6666666?
(b) 0.06?
(c) 0.1111111?

11 (i) Put the number 571 into your calculator.
Multiply it by 3 and write down the last digit in the display.
Multiply the new display by 3 and write down the last digit.
Repeat this until you can see a pattern in the last digits.
(ii) Repeat the process with 453 as the starting number.
(iii) Repeat the whole process with 2789 as the starting number.
(iv) The last three starting numbers have been odd numbers.
Find the last digit of an odd number which causes the pattern to change.

12 Katy has to calculate
$(14.4 - 3.6) \times (7.2 + 2.8)$.
What is the right answer?
Katy pressed one wrong button and got the result 180.
Which wrong button did she press?

2 decimals, percentages and fractions

DECIMALS ON THE NUMBER LINE

These lines are part of the number line.
Be careful! Each line has been magnified differently!
Write down the numbers indicated by the arrows.

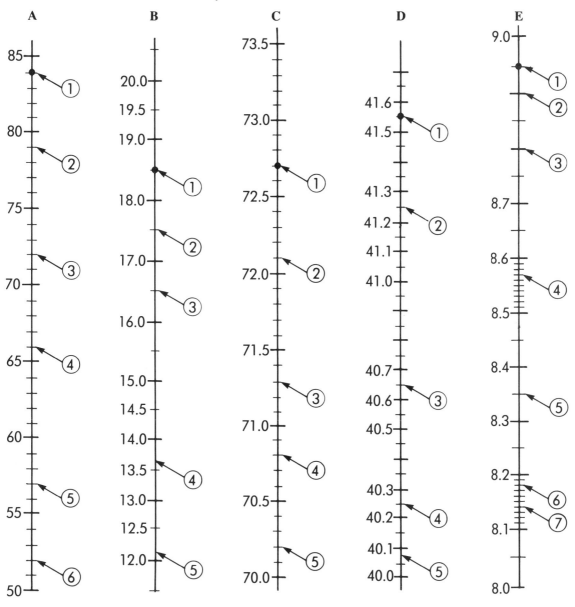

EXAMPLE

$567 = 5$ units $+ 6$ tenths $+ 7$ hundredths

$\frac{5}{8} = 0.625$

$9 + \frac{5}{8} = 9\frac{5}{8}$

$9 + 0.625 = 9.625$

$$3.5 \times 0.04 = 3.5 \times \frac{4}{100}$$
$$= 14 \div 100$$
$$= 0.14$$

A Calculate:

1 9.8×0.001
2 3.56×0.02
3 4.07×0.04
4 14.8×0.3
5 5.65×0.04
6 $96.3 \div 0.2$
7 $8.4 \div 0.2$
8 $0.86 \div 0.04$
9 $3.06 \div 0.6$
10 $624 \div 0.6$
11 0.48×0.03
12 0.03×1.05
13 0.07×2.53
14 5.06×0.5
15 0.4×5.06

B Calculate:

1 0.9×0.001
2 0.1×0.1
3 0.08×0.5
4 0.3×0.3
5 0.4×0.4
6 $0.9 \div 0.2$
7 $0.09 \div 0.3$
8 $0.08 \div 0.4$
9 $0.16 \div 0.004$
10 $0.3 \div 0.03$
11 0.02×0.02
12 0.5×0.5
13 0.07×0.07
14 0.8×0.5
15 0.05×0.06

C Write down the value of the 7 in:

1 317.4
2 6.7
3 54.7
4 34.17
5 715.6
6 3.75
7 201.073
8 57.8
9 0.07
10 0.107

D Write in figures:

1 two thousand and one
2 five thousand and twelve
3 fifteen thousand and thirty
4 three million
5 twelve hundred
6 two thousand and twenty-five
7 forty point eight
8 twenty-five thousand seven hundred
9 ten point nought three
10 two thousand six hundred and fifty

E
1 Write the next number in the sequence 27, 2.7, 0.27, . . .
2 Write the next number in the sequence 0.0015, 0.15, 15, . . .
3 Write the next number in the sequence 20, 2, 0.2, . . .
4 Write the next number in the sequence 4, 2, 1, 0.5, . . .
5 Write the next number in the sequence 80, 20, 5, . . .

EXAMPLE

$1\% = \frac{1}{100} = 0.01$ $5\% = \frac{5}{100} = \frac{1}{20} = 0.05$

$10\% = \frac{10}{100} = \frac{1}{10} = 0.1$ $25\% = \frac{25}{100} = \frac{1}{4} = 0.25$

A Express as a %

1 $\frac{3}{10}$ 6 $\frac{25}{75}$

2 $\frac{8}{16}$ 7 $\frac{9}{12}$

3 $\frac{5}{25}$ 8 $\frac{20}{30}$

4 $\frac{7}{10}$ 9 $\frac{25}{25}$

5 $\frac{45}{50}$ 10 $\frac{2}{5}$

B Express as a decimal

1 40%   6 84%

2 37%   7 29%

3 25%   8 70%

4 $12\frac{1}{2}\%$ 9 16%

5 60%   10 100%

C Express as a fraction

1 50%   6 30%

2 75%   7 25%

3 20%   8 $12\frac{1}{2}\%$

4 40%   9 $37\frac{1}{2}\%$

5 15%   10 60%

D Write down   (i) the fraction of each diagram which has been shaded.

           (ii) the percentage of each diagram which has been shaded.

           (iii) the percentage of each diagram which has not been shaded.

1
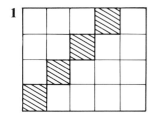

4
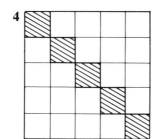

7
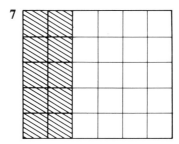

2
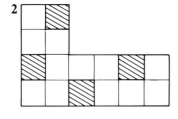

5
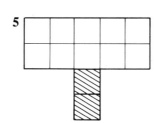

8
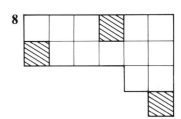

3
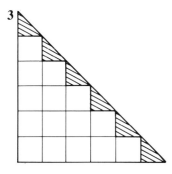

6
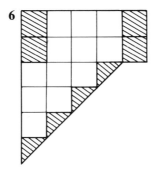

9
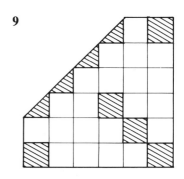

EXAMPLE

$1\% = \frac{1}{100}$

$5\% = \frac{5}{100} = \frac{1}{20}$

$25\% = \frac{25}{100} = \frac{1}{4}$

75% of £120 = $\frac{3}{4}$ of £120 = £90

2 days is $\frac{1}{5}$ of 10 days

so 2 days is 20% of 10 days

A Find

1 10% of £45
2 25% of £48
3 20% of £75
4 40% of £800
5 1% of £680
6 20% of 30 metres
7 25% of 36 metres
8 10% of 70 metres
9 5% of 9 metres
10 1% of 57 metres
11 10% of 580 cm
12 25% of 60 cm
13 30% of 450 cm
14 5% of 250 cm
15 2% of 65 cm
16 5% of 40 g
17 90% of 200 g
18 4% of 70 g
19 15% of 200 g
20 5% of 840 g

B Express as a percentage

1 40 out of 80
2 6 out of 24
3 50 out of 250
4 20 out of 60
5 6 out of 30
6 4 m out of 16 m
7 2 m out of 40 m
8 30 m out of 40 m
9 6 m out of 10 m
10 45 m out of 90 m
11 1 day out of 10 days
12 8 days out of 40 days
13 6 days out of 60 days
14 15 days out of 60 days
15 20 days out of 50 days
16 12 kg of 48 kg
17 6 kg of 200 kg
18 50 kg of 150 kg
19 60 kg of 90 kg
20 180 kg of 200 kg

C 1 Bill planted 250 lettuces on his allotment and lost 50. What percentage was that? He also planted 200 lettuces in his garden and lost 50 of them. What percentage did he lose in the garden? Which was the better place to grow lettuce?

2 In the summer a factory employed 200 girls. 16 were absent on August 10th. In the winter they only employed 50 girls and 3 were absent on December 1st. On which day was the larger percentage absent?

3 The number of people unemployed in Seltown fell from 2400 by 2%. What was the new total of people unemployed?

4 An employer promised his staff a 10% increase in wages. What will be the new wage of someone earning £170 a week? How much extra will a person get who was earning £65 per week?

5 The population of Coltown is 15 000 and it is expected to grow 5% in the next year. How many more people are expected to be living in the town?

6 The shopkeeper said all prices would be reduced by 20% in the sale. What would be the sale price of a coat marked £120?

7 Fred did two tests and scored 40 out of 50 on the first. His score on the second was 45 out of 60. Which was the better result?

16

EXAMPLE

15% of 60 = 0.15 × 60 = 9   60 decreased by 15% becomes 51

When 60 is increased by 15% it becomes 69

A

1 Increase 50 by 6%
2 Increase 40 by 10%
3 Increase 700 by 10%
4 Increase 64 by 25%
5 Increase 100 by 30%
6 Increase 80 cm by 5%
7 Increase 300 cm by 2%
8 Increase 450 cm by 20%
9 Increase 55 cm by 20%
10 Increase 800 cm by 50%
11 Decrease 600 tons by 10%
12 Decrease 40 tons by 25%
13 Decrease 70 tons by 3%
14 Decrease 150 tons by 30%
15 Decrease 2000 tons by 20%
16 Decrease 150 km by 2%
17 Decrease 100 km by 1%
18 Decrease 15 km by 50%
19 Decrease 40 km by 12.5%
20 Decrease 60 km by 5%

B Find the new price

1 Original price £60, price rise 10%
2 Original price £500, price rise 30%
3 Original price £400, price rise 25%
4 Original price £700, price rise 50%
5 Original price £48, price rise 25%
6 Original price £360, price rise 20%
7 Original price £300, price rise 30%
8 Original price £150, price rise 5%
9 Original price £60, price rise 7%
10 Original price £450, price rise 20%
11 Normal price £80, sale reduction 20%
12 Normal price £900, sale reduction 40%
13 Normal price £50, sale reduction 25%
14 Normal price £90, sale reduction 10%
15 Normal price £60, sale reduction 25%
16 Normal price £760, discount 5%
17 Normal price £860, discount 25%
18 Normal price £40, discount 75%
19 Normal price £600, discount 15%
20 Normal price £90, discount 75%

C The price for this house is made up of:

Builder's profit	10%	Materials	35%
Labour costs	40%	Other costs	

1 How much profit does the builder expect to make?
2 How much were the labour costs?
3 How much did the materials cost?
4 What percentage of the selling price were the other costs?

FOR SALE £68 000

D The workforce of a small company is:

1 manager paid monthly	£18 000 p.a.
7 staff paid weekly	£7800 p.a.
3 juniors paid weekly	£3.50 per hour, overtime £5 per hour

1 Each junior works 30 hours a week. How much is the total weekly pay for a junior before any deductions are made?
2 How much is the weekly total before deductions for each of the staff?
3 The staff are given a 10% pay increase. What does the annual salary become?
4 The manager is given a 7% increase. By how much does his annual salary increase? How much extra is this per month?
5 One junior worked 30 hours and then 4 hours overtime. What was his total pay for that week?

17

1 Ann works in a large store. The staff are allowed 8% discount on any purchases they make in the store. She wants a jacket priced £90 and towels costing £40. How much will Ann have to pay for them?

2

Special offer!!

5% reduction
for prompt payment

Mr Harris has a bill for £3500. How much would he save by paying promptly?

3

PRETTY HOMES

To Mrs Jones
14 Sunny Way

	Dining Set Tax @ 15%	£780
	Total	
2% discount for cash To pay		

Mrs Jones can pay for her furniture as soon as it is delivered.
Calculate the tax and full price she would have to pay.
How much does she save by paying cash?
How much does she pay?

4 The cost of having a wall repaired was £350.
A tax of 15% was added to the cost.
What was the total to be paid?

5 A salesman's pay is made up like this:
  basic pay: £120 per week;
  first £200 worth of sales: no commission;
  sales above £200: 5% commission.
In a week when he makes £150 worth of sales how much is his pay?
When he makes £600 worth of sales, how much is his commission?
In a week when he makes £1000 worth of sales, what is his total pay?

6 Mr Edwards paid £9500 for a new car. The depreciation on a new car is about 20% in the first year. About how much is his car worth when it is 1 year old?

7

£4357 on the road
or 30% deposit
and £96.46 for 36 months

How much is the deposit?
What is the total to pay on HP?
How much extra does this HP system cost?
Is the cost of HP about 10%, 15%, 20% or 30% of the cash price?

8 The salaries of a company cost £320 000 per annum.
25% of the total is paid to the managers. How much is this?
The rest is paid to the shopfloor workers. How much is this?
Salaries negotiations gained a 7% increase for the managers and 8% for the shopfloor workers. Calculate the total increase in the salaries paid by the company.

A Two pupils made a survey of the ways pupils travelled to East Road School.

Age 16	Bus	Car	Walk	Cycle
16	///// /	/// ////	/// /	///// ///
15	///	///	//	//
12	/// //	// //	//	//
11		/////		

Age	Bus	Car	Walk	Cycle
16	xxxx xxx	xxxx x	xxxx xx	xxxx
15	xx			x
12	xxxx	xxxx xxx	xx	xxx
11	x	xxxx	x	

They started to combine their results in one table. Copy and complete it.

Age group	Bus	Car	Walk	Cycle	Total
15's and 16's	18	15	12	15	60
% of age group	30%	25%			
11's and 12's					
% of age group					

Discuss the different patterns in the two age groups.

There are 250 pupils at the school aged 15 or 16. If this group of 15's and 16's is typical of all the 15's and 16's in East Road School then how many of them cycle to school?

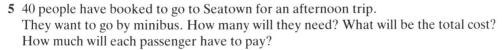

B

K Transport	Number of buses	Number of seats	Hire charges to Seatown Half day/evening	Whole day
Minibus		20	£50	£90
Single decker			£65	£110
Double decker			£80	£150

K Transport have three kinds of bus in their fleet.
They have 30 single deckers which are 50% of their fleet.

1 How many buses are in the fleet?
2 Copy and complete the above table.

Buses can be hired for outings.

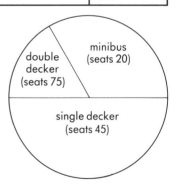

3 Which buses should be booked to take 80 people on an outing?

4 What would be booked to take 70 people by bus?

5 40 people have booked to go to Seatown for an afternoon trip.
They want to go by minibus. How many will they need? What will be the total cost?
How much will each passenger have to pay?

6 What is the cheapest way of taking 40 people to Seatown by K Transport?
How much should each passenger be charged?

EXAMPLE

$\frac{1}{3} \times 4 = 1\frac{1}{3}$ $\frac{3}{8} \times 2 = \frac{6}{8} = \frac{3}{4}$ $\frac{1}{10} \div 2 = \frac{1}{20}$

$\frac{6}{7} \div 2 = \frac{3}{7}$

A
1. $1\frac{1}{2} \times 5$
2. $\frac{1}{4} \times 6$
3. $\frac{1}{3} \times 3$
4. $\frac{1}{5} \times 6$
5. $\frac{1}{2} \times 3$
6. $\frac{1}{10} \times 20$
7. $\frac{1}{5} \times 12$
8. $\frac{1}{3} \times 5$
9. $\frac{1}{2} \times 6$
10. $\frac{1}{4} \times 8$
11. $1\frac{1}{2} \times 3$
12. $2\frac{1}{4} \times 4$
13. $5\frac{1}{2} \times 2$
14. $3\frac{1}{5} \times 4$
15. $7\frac{1}{4} \times 2$
16. $10\frac{1}{2} \times 3$
17. $1\frac{1}{3} \times 6$
18. $1\frac{1}{4} \times 8$
19. $3\frac{1}{3} \times 4$
20. $6\frac{1}{5} \times 2$

B
1. $1\frac{2}{3} \times 2$
2. $\frac{4}{5} \times 3$
3. $\frac{3}{5} \times 2$
4. $\frac{3}{4} \times 3$
5. $\frac{3}{5} \times 5$
6. $\frac{3}{8} \times 3$
7. $\frac{5}{6} \times 2$
8. $\frac{7}{8} \times 3$
9. $\frac{4}{5} \times 5$
10. $\frac{3}{10} \times 4$
11. $1\frac{1}{3} \times 3$
12. $1\frac{2}{5} \times 2$
13. $1\frac{3}{5} \times 2$
14. $2\frac{3}{4} \times 2$
15. $3\frac{4}{5} \times 2$
16. $5\frac{3}{7} \times 2$
17. $4 \times \frac{7}{8}$
18. $4 \times 1\frac{7}{8}$
19. $6 \times 3\frac{2}{5}$
20. $5 \times 2\frac{2}{3}$

C
1. $1\frac{1}{5} \div 2$
2. $\frac{1}{4} \div 2$
3. $\frac{3}{4} \div 2$
4. $\frac{4}{5} \div 2$
5. $\frac{1}{2} \div 3$
6. $\frac{1}{4} \div 3$
7. $\frac{3}{4} \div 3$
8. $\frac{2}{5} \div 4$
9. $\frac{2}{3} \div 4$
10. $1\frac{2}{3} \div 5$
11. $2\frac{1}{2} \div 5$
12. $3\frac{3}{5} \div 3$
13. $4\frac{6}{7} \div 2$
14. $8\frac{4}{9} \div 4$
15. $4\frac{3}{8} \div 2$
16. $5\frac{1}{2} \div 3$
17. $10\frac{3}{8} \div 5$
18. $6\frac{3}{10} \div 3$
19. $1\frac{1}{5} \div 3$
20. $5\frac{1}{3} \div 4$

D Use fractions to label the points on these parts of the number line:

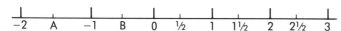

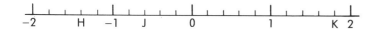

EXAMPLE

$\frac{2}{3} + \frac{1}{5} = \frac{10}{15} + \frac{3}{15} = \frac{13}{15}$

A Calculate:

1. $\frac{1}{2} + 1\frac{1}{2}$
2. $1\frac{1}{4} + \frac{1}{2}$
3. $\frac{1}{3} + \frac{1}{6}$
4. $\frac{2}{5} + \frac{1}{10}$
5. $\frac{2}{3} + \frac{1}{6}$
6. $3\frac{1}{2} + \frac{2}{5}$
7. $\frac{1}{4} + \frac{1}{6}$
8. $4\frac{2}{3} + \frac{1}{5}$
9. $\frac{3}{4} + 6\frac{1}{5}$
10. $\frac{1}{3} + \frac{1}{2}$
11. $1\frac{3}{5} + \frac{1}{2}$
12. $2\frac{1}{4} + \frac{5}{8}$
13. $5\frac{3}{5} + \frac{3}{8}$
14. $2\frac{1}{2} - \frac{1}{4}$
15. $1\frac{5}{9} - \frac{1}{3}$
16. $4\frac{3}{8} - \frac{1}{4}$
17. $10\frac{1}{2} - 8\frac{3}{4}$
18. $6\frac{5}{8} - 3\frac{7}{8}$
19. $5\frac{2}{3} + 1\frac{3}{4}$
20. $8\frac{5}{6} + 2\frac{1}{3}$

B Express as decimals

1. $3\frac{1}{2}$
2. $12\frac{1}{2}$
3. $5\frac{3}{4}$
4. $4\frac{1}{8}$
5. $7\frac{3}{10}$
6. $5\frac{4}{5}$
7. $2\frac{1}{4}$
8. $\frac{7}{8}$
9. $3\frac{1}{6}$
10. $\frac{1}{40}$
11. $\frac{3}{40}$
12. $4\frac{3}{40}$
13. $\frac{1}{20}$
14. $\frac{3}{20}$
15. $\frac{7}{20}$
16. $7\frac{19}{20}$
17. $\frac{1}{50}$
18. $\frac{7}{50}$
19. $9\frac{49}{50}$
20. $6\frac{7}{8}$

C Find

1. $\frac{1}{5}$ of £10
2. $\frac{1}{4}$ of £8
3. $\frac{1}{2}$ of £9
4. $\frac{1}{3}$ of £3.60
5. $\frac{1}{5}$ of £6
6. $\frac{1}{5}$ of 1 kilometre
7. $\frac{1}{4}$ of 2 kilometres
8. $\frac{1}{5}$ of 600 g
9. $\frac{1}{3}$ of 450 g
10. $\frac{1}{3}$ of 63 cm
11. $\frac{2}{5}$ of £10.60
12. $\frac{3}{4}$ of 560 g
13. $\frac{7}{20}$ of £4.80
14. $\frac{7}{8}$ of £10.40
15. $\frac{5}{9}$ of 18 kg
16. $\frac{2}{3}$ of 15 metres
17. $\frac{4}{5}$ of 2 hours
18. $\frac{3}{10}$ of 1 hour
19. $\frac{3}{4}$ of 20 minutes
20. $\frac{1}{3}$ of 1 day

D 1 Mary and Kathy work for a basic wage of £4 per hour.

The rate of pay is increased by half for overtime. How much is this?

Why is it sometimes said that this is 'time and a half'?

In a week when Mary works 30 hours and then does 4 hours overtime how much will she have earned?

Working nights is paid at overtime rates.

Kathy does 5 hours at the basic rate and then 20 hours on nights. How much does she earn?

2 The sports commentator said the horse won by $1\frac{1}{2}$ lengths.
Was that about 3 yards, 15 metres or 1.5 metres?

directed numbers

THE NUMBER LINE

EXAMPLE

$-2 + 3 = 1$

$-5 + 3 = -2$

$2 - 3 = -1$

$2 < 5$

$-1 < 3$

$-4 < -2$

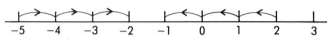

Moving $\xrightarrow{+}$ up the number line is adding, and moving $\xleftarrow{-}$ is subtracting.

A Copy and complete:

	Start	Move	Finish	equation
1	1	+3	4	$1 + 3 = 4$
2	3	+4	7	$3 + 4 =$
3	−2	+3		$-2 + 3 =$
4	−1	+5		
5	−3	+5		
6	−5	+3		
7	−7	+2		
8	−4	+5		
9	−2	+2		
10	−5	+4		

	Start	Move	Finish	equation
11	3	−2	1	$3 - 2 = 1$
12	5	−3	2	
13	2	−3		
14	3	−4		
15	5	−3		
16	−2	−1		
17	−3	−3		
18	0	−3		
19	6	−6		
20	−1	−4		

B Use $<$, $>$ or $=$ to make a true statement.

1	3	5
2	−1	4
3	−2	0
4	−3	−3
5	2	−4
6	3	−3
7	5	−7
8	−2	−3
9	4	2
10	−1	−5

11	−5	9
12	−4	−3
13	2	−2
14	−1	0
15	5	−5
16	−7	−6
17	−2	−1
18	8	−9
19	−3	−5
20	−4	−6

C Make a list of the values of x which make the statement true. In these examples, x is always a whole number.

1 $2 < x < 5$	**11** $-1 < x < 3$
2 $-3 < x < 2$	**12** $-4 < x < -1$
3 $3 < x < 7$	**13** $-1 < x < 2$
4 $-2 < x < 3$	**14** $-5 < x < -2$
5 $0 < x \leqslant 3$	**15** $-7 < x < -4$
6 $3 < x < 6$	**16** $-4 \leqslant x < -3$
7 $-3 < x < 1$	**17** $5 < x < 8$
8 $-1 < x < 1$	**18** $-8 < x < -6$
9 $0 < x < 2$	**19** $-2 < x < 0$
10 $8 < x \leqslant 9$	**20** $-9 < x < -7$

PUSH AND LEAN

Two men called PUSH and LEAN are at work on the number line.
They have to move a box to and fro.

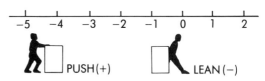

LEAN got his name because he is lazy and always leans, but it does make his box move.
The moves they make are coded and follow certain rules:
Adding always means facing →   Subtracting always means facing ←

A PUSH starts work first. All his moves are written (+ ..)

   Adding ——→      Subtracting ←——

The codes for these moves are:

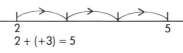

$$2 + (+3) = 5$$

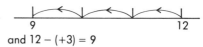

and $12 - (+3) = 9$

Write the code for these moves:

1    2 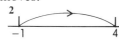   3    4

B LEAN's moves make the box move in the opposite direction to PUSH, so all his moves are written (− ..)
LEAN is facing forwards so it is add: $6 + (-2) = 4$

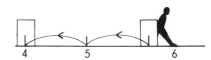

The codes for these moves are:

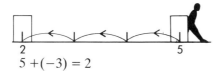

$$5 + (-3) = 2$$

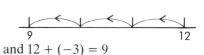

and $12 + (-3) = 9$

Write the code for these moves:

1 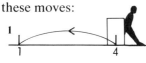   2    3    4

C For subtraction, LEAN faces the other way, just as PUSH did. LEAN's moves are always written (− ..) whichever way he is facing. LEAN makes the box move (−2) places from 3. It goes to 5: The sum is $3 - (-2) = 5$
Draw the arrows on each diagram to find the answers. Remember LEAN is subtracting.

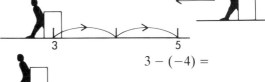

$0 - (-3) =$ $3 - (-4) =$

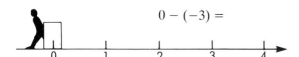

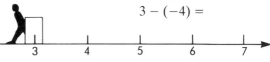

DISTANCE APART

EXAMPLE

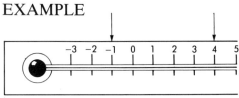

−1 °C is 5° below 4 °C
4 − (−1) = 5

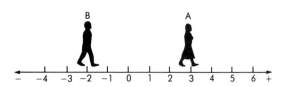

B is 5 steps behind A
3 − (−2) = 5

A Find the difference between these temperatures:

1 −2 °C, 4 °C	**11** −1 °C, −2 °C
2 −2 °C, 6 °C	**12** −3 °C, −1 °C
3 −1 °C, 1 °C	**13** −4 °C, −2 °C
4 −2 °C, 3 °C	**14** 5 °C, 3 °C
5 −2 °C, 6 °C	**15** 7 °C, −1 °C
6 −3 °C, 4 °C	**16** 10 °F, 12 °F
7 1 °C, 8 °C	**17** 8 °F, 3 °F
8 6 °C, 2 °C	**18** −2 °F, 10 °F
9 5 °C, −1 °C	**19** 12 °F, −1 °F
10 3 °C, −2 °C	**20** 16 °F, 20 °F

B Calculate:

1 3 − (+1)	**11** 4 − (−5)
2 5 − (+2)	**12** 6 − (+2)
3 6 − (−3)	**13** 0 − (−4)
4 7 − (+3)	**14** 3 − (−3)
5 9 − (−1)	**15** 2 − (−3)
6 5 − (−4)	**16** 1 − (−4)
7 10 − (+7)	**17** 2 − (+1)
8 0 − (−3)	**18** 5 − (−6)
9 0 − (−1)	**19** 4 − (−5)
10 2 − (−2)	**20** 3 − (−4)

C Martown is liable to flooding when the river water rises above a certain level.
The board has been marked so that 0 is the level at which warnings are to be given.
The readings for a three week period are shown in the table:

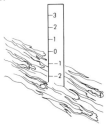

	Mon	Tues	Wed	Thur	Fri	Sat	Sun
Week 1	−1.9	−2.0	−2.0	−1.8	−1.4	−0.8	0
Week 2	0.5	0.8	0.8	0.5	0	−0.4	−0.6
Week 3	−0.7	−0.8	−1.4	−1.9	−2.0	−2.0	−2.0

1 When was the flood warning given?
2 When was the biggest rise? How much was it?
3 When did the water stop rising?
4 How long was the water above the warning level?

5 What was the highest reading recorded?
6 What seems to be the usual level?
7 Which graph represents these readings?

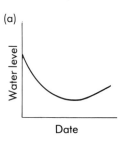
(a)

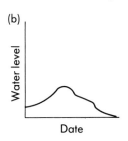
(b)

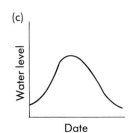
(c)

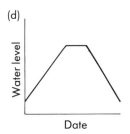
(d)

4 money, rates of charging

ONE THOUSAND POUNDS INVESTED

This graph was produced by a company wishing to show how £1000 had changed in value since it was invested with them in June 1978.

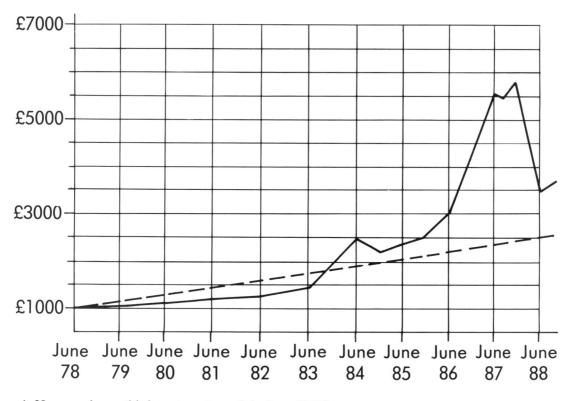

1 How much was this investment worth in June 1988?
2 How much had its value increased in 10 years?
3 What was the average growth in 1 year?
4 When was the greatest value and how much was this?
5 When was the investment growing most quickly?
6 In which periods did the value of the investment fall?
7 When did the biggest fall in value occur?
8 How much was the investment worth in June 1983?
9 If the graph is read carelessly, it could appear that the value had doubled by June 1983. How could this mistake be made?
   How could the graph be drawn to avoid the risk of a mistake?
10 Did the value of the investment ever reach £7000?
11 The broken line shows how £1000 would have grown if it had been in a building society account. When would this have been a better way of investing the money?

ADVERTISING

1 Read as much as you can of the advertisements at the side.

The advertisement about freezers is one column wide. Measure its depth.

It is a trade advertisement. Look at the table of charges to find how much it will cost for one day.

How much will it cost for three days?

There is a lower rate for four or five days. How much will five days cost?

2 Measure the depth of the Crewe Theatre advertisement.

How much will it cost for one day?

How much will it cost for six days?

3 Look at the advertisement for the lounge suite. What is the price wanted?

How many lines are in the advertisement?

How much will it cost for one day?

How much will it cost for three days? (N.B. three days for price of two)

4 Look at the advertisement for the knitting machine.

How many lines does it use?

How much will it cost for one day?

How much will it cost for two days?

How much will it cost for three days?

5 An advertisement should appear on Friday. Which day must it be taken to the newspaper office?

FREE

Cutting – Packing & Delivery
TO
DEEP FREEZER OWNERS
ΊE KILLED MEATS
h Lamb (whole)
9p lb.
ump (incl.
50p lb

ELECTION
MEATS

E 3765

Crewe
CREWE THEATRE   CREWE 5
Tonight at 7.50: Sat. at 8 pm:

THE IMPORTANCE OF BEING EARNEST

Monday only:

MANCHESTER CAMERATA

BEAUTIFUL UNUSED Gold Dralon
Lounge Suite Comprising three Seater
Settee One swivel Rocker Chair One
Armchair and Footstool All matching.
Accept £135 the lot. Can arrange
delivery – 670 Abbey Lane, Ro
nts Tel R

KNITMASTER knitting machine
wanted, any model considered. Ring
Stockton

ADVERTISING CHARGES IN THE EVENING GAZETTE	
Minimum order:	Private advertisers . . . 2 lines for 1 evening Display (in a frame) . . . 3 cm deep, 1 column wide
Copy deadline:	4.30 pm two days before publication
Trade and business rates:	
1, 2, 3 days 40p per line per day 4, 5 days 35p per line per day	£2.05 per column cm per day £1.70 per column cm per day
Private advertisers' rates:	
Items for sale over £50.23p per line per day Items for sale under £50.12p per line per day (3 consecutive days for the price of 2)	
£1.25 per column cm per day £0.60 per column cm per day	
Entertainments:	Display £2.20 per cm column per day

ESTIMATING COSTS

A Use $<$, $>$ or $=$ to make true statements. Do not use a calculator.

1 £19.80 × 2   £40
2 £29.99 × 5   £150
3 £123 × 40   £5000
4 £592 × 50   £30 000
5 £47.81 × 6   £300

6 £25.70 × 5   £125
7 £35.80 × 4   £144
8 £67.83 × 20   £1300
9 £50.27 × 41   £2000
10 £102.39 × 5   £500

11 £48 × 20   £1000
12 £95 × 300   £30 000
13 £730 × 50   £35 000
14 £580 × 60   £40 000
15 £700 × 20   £14 000

B **1** Mrs Harris says her cat weighs 4 kg and it costs her more than £200 a year to feed it. About how much is this each week?

2 Mrs Brown wants a new carpet for her lounge.

£12.55 per sq.m Long Life

£15.55 per sq.m Extra Quality

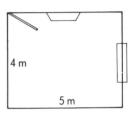

4 m

5 m

The salesman says it will cost about £30 extra to have the carpet fitted. She likes the Extra Quality carpet. How much will it cost to have it fitted in this lounge (to the nearest £10)? If she had chosen the Long Life carpet, how much less would it cost her?

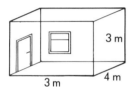

3 m

3 m   4 m

£4.99 2.5 litres covers 40 sq.m WHITE

£4.99 2.5 litres covers 40 sq.m GREY

3 The ceiling of this room needs one coat of white emulsion paint. The walls are to be grey and will need two coats of paint. How much will it cost to buy the paint?

4 Gas is sold in units called therms. The consumer has to pay a basic charge each quarter. The cost of the units he has used is added to the basic charge.

This is a gas tariff:

Standing charge per quarter	£3.50
First 40 therms per quarter	56.7 p
Further therms per quarter	41.0 p

Mr Stone always uses more than 40 therms in a quarter.

Make a statement about his quarterly bill.

Copy and complete his quarterly statement:

	Mr Stone
Previous meter reading 0 2 3 5	
Present meter reading 0 3 7 5	
Quarterly standing charge	£ ....
First 40 therms @ 56.7p	£ ....
therms @ 41.0p	£ ....
Amount to pay	£ ....

RATES

A 1

BEST RATE	HIGH INTEREST!	GILT FUND
10¾%	10.7%	10.69%
High Return	Super rate	Unbeatable!

Which of these savings schemes is offering the highest return on an investment?
What is the most interest that £500 could earn in 1 year on one of these schemes?

2

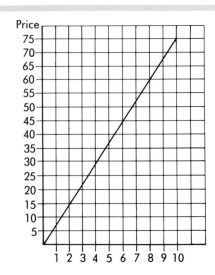

50 litres/minute

This new garden pond is to be filled from a hose. It is estimated that it will need 25 000 litres to fill it. How long will this take?
If the water is turned on at 8.30 am when should it be checked?

3 1 litre of fluid has to be given to a patient. It has to be given at a steady rate over a period of 8 hours.
If it starts at 10.00 am when will it finish?
How much should be given in 1 hour?
How many ml will be given per minute?
The patient is given the fluid through a tube in drops. There are 20 drops in 1 ml.
How many drops is this per minute?
The nurse sets the machine to drip at 42 drops per minute. Will it finish before or after 6.00 pm?

4 The petrol tank of Mr Green's car holds 11 gallons. The gauge shows it is half full. About how much petrol is in the tank?
Mr Green has to travel 210 miles and he thinks the car does about 35 miles to the gallon. Will he need more petrol?

5 The Family Size shampoo costs 85p per litre. There is a 200 ml bottle of the same shampoo which costs 39p. Which is the better buy, and why?

B This graph can be used as a calculator. It shows that if 1 unit costs 7.5p, 10 cost 75p

1 Find the cost of 8 units at 7.5p each.

2 Find the cost of 5 units at £7.50 each.

3 Find the cost of 3 units at £7.50 each.

4 How many items at 7.5p each can be bought for 50p?

5 Find the cost of 6 kg at 7.5p per kg.

6 How many things costing 7.5p each can be bought for 40p?

7 How many things costing £7.50 each can be bought for £70?

EXAMPLE

The ratio $12 : 15 = 3 \times 4 : 3 \times 5$
$$= 4 : 5$$

The ratio of $4\,\mathrm{kg} : 6\,\mathrm{kg} = 2 : 3$

£20 divided in the ratio $2 : 3$ is split into 5 equal parts. 2 parts is $2 \times £4 = £8$ and 3 parts is $3 \times £4 = £12$

A Simplify these ratios:

1. $3 : 6$
2. $12 : 4$
3. $4 : 12$
4. $15 : 25$
5. $25 : 100$
6. $24 : 16$
7. $6 : 18$
8. $20 : 50$
9. $14 : 21$
10. $18 : 45$

B Simplify these ratios:

1. $5\,\mathrm{kg} : 15\,\mathrm{kg}$
2. $40\,\mathrm{cm} : 1\,\mathrm{m}$
3. $1\,\mathrm{m} : 25\,\mathrm{cm}$
4. $250\,\mathrm{g} : 1\,\mathrm{kg}$
5. $500\,\mathrm{g} : 2\,\mathrm{kg}$
6. $600\,\mathrm{m} : 1\,\mathrm{km}$
7. $5\,\text{litres} : 3\,\text{litres}$
8. $75\,\mathrm{cl} : 1\,\text{litre}$
9. $25\,\text{yards} : 50\,\text{yards}$
10. $4\,\mathrm{lb} : 12\,\mathrm{lb}$

C Divide:

1. £16 in the ratio $3 : 5$
2. £100 in the ratio $3 : 7$
3. £60 in the ratio $5 : 7$
4. 50 yards in the ratio $1 : 4$
5. 10 inches in the ratio $2 : 3$
6. 28 tons in the ratio $4 : 3$
7. 56 lb in the ratio $3 : 5$
8. 64 cm in the ratio $5 : 3$
9. 500 g in the ratio $3 : 7$
10. 1 kg in the ratio $2 : 3$

D

1. Make an accurate drawing of a line AB so that $AB = 12.0\,\mathrm{cm}$
   Mark the point P on AB so that $AP : PB = 1 : 3$
2. Make an accurate drawing of a line XY so that $XY = 10.0\,\mathrm{cm}$
   Mark the point P on AB so that $XP : PY = 2 : 3$
3. Make an accurate drawing of a line AB so that $AB = 12.5\,\mathrm{cm}$
   Mark the point P on AB so that $AP : PB = 2 : 3$
4. Make an accurate drawing of a line CD so that $CD = 10.0\,\mathrm{cm}$
   Mark the point P on AB so that $CP : PD = 1 : 4$
5. Make an accurate drawing of a line AB so that $AB = 9.6\,\mathrm{cm}$
   Mark the point P on AB so that $AP : PB = 1 : 2$

E

1. There is £120 prize money to be shared in the ratio $2 : 1$ between the people in first and second places. How much does each get?

2. Sam and Henry have agreed to share the profits from their market stall in the ratio $3 : 5$. On the day when they made £240 profit, how much was Henry's share?

3. In his will, Mr Gray has given money to his family. He wants the remainder sent to the Tabby Cat Lovers Society and the Goldfish Society. There is £480 for these two and it is to be shared in the ratio $2 : 1$. How much will each society get?

4. Ann uses a recipe for scones in which the ratio of margarine to flour is $1 : 4$. She wants to use 30 g of margarine. How much flour will she need?

5. A company spends £35 000 on wages every month. It is split between the office staff and sales staff in the ratio $3 : 4$. In June, how much do they pay to sales staff?

HOW MUCH WILL IT COST?

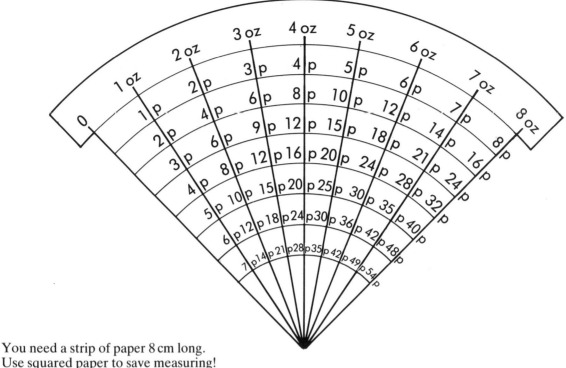

You need a strip of paper 8 cm long.
Use squared paper to save measuring!
Label it like this:

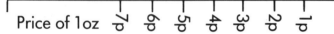

Price of 1oz   7p 6p 5p 4p 3p 2p 1p

To find the cost of 3 oz put the strip on the
chart above like this:

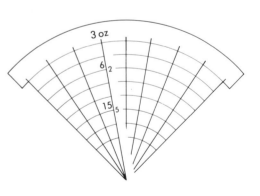

The '2p for one ounce' on the strip is against 6p.

1 Look at 5p on the strip. How much will 3 oz cost?

2 To find the cost of 5 oz, turn the top of the
   strip till the black edge is against 5 oz. Check
   the corner is still on the dot.
   If 1 oz costs 4p, how much does the chart say
   5 oz will cost?

Use the chart and strip to do these:

3 1 oz costs 4p. Find the cost of 6 oz.
4 1 oz costs 7p. Find the cost of 5 oz.
5 1 oz costs 6p. Find the cost of 7 oz.
6 6 oz cost 30p. How much does 1 oz cost?
7 7 oz cost 35p. How much does 1 oz cost?

Make up some more sums to do on the chart.

EXCHANGE RATES

EXAMPLE
The exchange rate changes from day to day. On this day £1 = 3 marks   10.5 francs = £1

$£5 = 3 × 5$ marks   21 francs = £(21 ÷ 10.5)

= 15 marks   = £2

Denmark	12 kroner	France	10.5 francs	Germany	3 marks
Italy	2300 lira	Portugal	250 escudos	Spain	200 pesetas

Using these tourist rates for sterling, make the following conversions:

A
1 £2 to kroner
2 £4 to marks
3 £10 to lira
4 £5 to pesetas
5 £50 to francs
6 £100 to escudos
7 £30 to pesetas
8 £200 to francs
9 £500 to escudos
10 £70 to marks
11 £25 to kroner
12 £50 to lira
13 £60 to marks
14 £10 to kroner
15 £30 to escudos
16 £100 to pesetas
17 £40 to francs
18 £50 to marks
19 £20 to lira
20 £100 to marks

B
1 500 escudos to £
2 36 marks to £
3 105 francs to £
4 1000 pesetas to £
5 300 marks to £
6 210 francs to £
7 1500 pesetas to £
8 3600 pesetas to £
9 1500 escudos to £
10 500 kroner to £
11 4000 pesetas to £
12 500 marks to £
13 5000 lira to £
14 50 francs to £
15 70 francs to £
16 100 kroner to £
17 60 kroner to £
18 10 000 lira to £
19 900 escudos to £
20 6000 pesetas to £

C
1 50 kroner to nearest £
2 100 marks to nearest £
3 70 kroner to nearest £
4 5000 lira to nearest £
5 100 francs to nearest £
6 2000 escudos to nearest £
7 10 000 lira to nearest £
8 1200 kroner to nearest £
9 22 francs to nearest £
10 7500 pesetas to nearest £
11 45 kroner to nearest £
12 75 francs to nearest £
13 65 marks to nearest £
14 96 francs to nearest £
15 6000 lira to nearest £
16 1000 kroner to nearest £
17 700 francs to nearest £
18 65 marks to nearest £
19 57 francs to nearest £
20 27 kroner to nearest £

D Use the exchange rates above for this section.

1 Which is greater in value, 21 francs or 7 marks?
2 Which is of less value, 1250 escudos or 1100 pesetas?
3 After my holiday I had 3000 escudos left. How much is that in sterling?
4 I have 100 marks. How much is that in English currency?
5 In a Danish shop I bought 3 books costing 25 kroner each and a picture which cost 130 kroner. How much should I have left out of 250 kroner?
6 In Italy I bought 5 scarves costing 25 000 lira each. How much change should I have had from 200 000 lira?
7 I shall be in Spain for 8 days and I want to allow myself £20 a day spending money. How many pesetas will I need?
8 Ken was in Germany for 3 days and bought himself 2 shirts costing 52 marks each and a tie for 33 marks. How much change did he have out of 150 marks?
9 At the end of a 10 day holiday in Portugal I had 8000 escudos left. How much was that in sterling?
10 Which is greater in value, 51 marks or 170 francs?

5 time

WATCHING TELEVISION

A 'Tops' starts at 7.20 pm
and finishes at 8.00 pm

so it lasts

```
   8.00
-  7.20
-------
  40 minutes
```

How long do these programmes last?

1	Sports News	8.00–8.30	**6**	Little Princess	5.15–5.40
2	Play Time	4.00–4.25	**7**	Midweek	10.50–11.30
3	News	6.00–6.55	**8**	Showtime	9.30–10.15
4	Lunch Hour	1.00–1.45	**9**	Film Night	10.35–12.00
5	Panto mania	8.10 9.00	**10**	Monsters	3.45–4.30

B Newsday starts at 7.30
and lasts 15 minutes

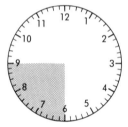

```
        7.30
     +    15
     -------
        7.45
```

It finishes at 7.45

What time do these programmes finish?

1	Nine o'Clock News	starts 9.00, lasts 25 minutes	**6**	The Workers	starts 4.25, lasts 5 minutes
2	Man at Work	starts 3.00, lasts 30 minutes	**7**	Teatime	starts 4.30, lasts 15 minutes
3	Call My Name	starts 9.00, lasts 25 minutes	**8**	Boyo	starts 7.20, lasts 50 minutes
4	Chigley	starts 1.45, lasts 20 minutes	**9**	Schools	starts 9.30, lasts 32 minutes
5	Look Here	starts 6.00, lasts 50 minutes	**10**	Taste of Britain	starts 7.45, lasts 25 minutes

C A class of children was asked to keep a
record of the number of hours they spent
watching TV on Saturday.

1 What was the longest time spent watching?
2 How many watched for 7 hours?
3 How many didn't watch at all?
4 How many children are in the class?
5 What was the most popular watching time?
6 What is the total of all the hours spent watching?
7 What was the average watching time per child?

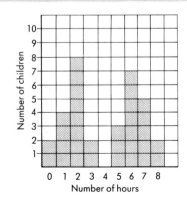

ARRIVING EARLY

EXAMPLE
The clock shows 2 minutes past 7
The time can be written 7.02

A Write the time shown by these clocks using past or to the hour:

1　　　　2　　　　3　　　　4　　　　5

B Write the time shown by these clocks using numbers but no words:

1　　　　2　　　　3　　　　4　　　　5

C Write the time 10 minutes before these times:

1 10.05 am	**3** 4.30 pm	**5** midday	**7** quarter past 10	**9** twenty to 4
2 quarter to 11	**4** half past 6	**6** 5 past 3	**8** 9.00 am	**10** 8.05 am

THE DOCTOR'S APPOINTMENTS

D Opposite is a page from a hospital appointment book:
  1 How many patients did the doctor expect to see?
    Do you think all of them will attend?
  2 The doctor sees each patient for five minutes.
    He has a 20 minute coffee break.
    He starts promptly at 9.40 am. When should he finish?
  3 Why do you think two patients are booked for 9.40 am?
  4 Copy and complete the chart below for all the patients.
    Remember each patient is with the doctor
    for just five minutes.

9.40 am	Mr A	Mrs B
9.45 am	Mr C	Mr D
9.50 am	Mrs E	
10.00 am	Miss F	
10.10 am	Mrs G	Mrs H
10.20 am	Miss J	
10.25 am	Mr K	
10.30 am	Mrs L	Mrs M
11.00 am	Mrs N	Mrs P
11.10 am	Mrs Q	Mrs R
11.20 am	Mrs S	Mrs T
11.30 am	Miss U	Mr V

  5 Which patients are seen at
    exactly the right time?
  6 How many patients have to wait
    more than five minutes?
  7 What is the longest wait a
    patient will have?
  8 If both Mrs N and Mrs P do not
    attend, how long will the doctor
    waste?

	Time on his card	Minutes late
Mr A is seen at 9.40 am	9.40 am	0
Mrs B is seen at 9.45 am	9.40 am	5
Mr C is seen at 9.50 am	9.45 am	5
Mr D is seen at 9.55 am		
Mrs E is seen at 10.00 am		

33

HOW OLD IS IT?

Manufacturers use a code to put the date of production on food cans.

A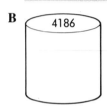
26588

26588 has to be split up like this: 26 5 88.
May is the 5th month, so this can was dated 26th May 1988.

Use this code to find the dates represented by these numbers:

1 18387   **2** 30976   **3** 25488   **4** 06589   **5** 07179

B
4186

How do you know this can has been dated with a different code?

This code does not mention the month, but it has to be thought out like this: The days are counted from the beginning of the year, so 1987 means 19 days after the start of the year, i.e. 19th January, and the last two digits give the year. Thus 1987 means 19th January 1987. The 10th February is 41st day of the year (January has 31 days + 10). So 10th February 1986 is written 4186.

How would you write 15th February 1986 in this code?

January 31 days   February 28 days   March 31 days   April 30 days   May 31 days
June 30 days   July 31 days   August 31 days etc.
Continue this: 1st Feb. is 32nd day, 1st Mar. is 60th day, 1st Apr. is . . .

Write these dates in this code:

1 3rd March 1956   **3** 20th May 1967   **5** 10th August 1975
2 4th April 1989   **4** 23 February 1990   **6** 5th September 1990

In this code, what dates do these code numbers represent?

1 7391   **2** 6886   **3** 9777   **4** 10587   **5** 36589   **6** 20085

C
86B21   87E14   78H29
92K15

This collection of cans comes from one manufacturer. To read this code try splitting up the numbers and letters like this: 86 B 21. Write the codes on each can in a list, one under the other and then ask yourself some of these questions:

Could the first or last pair of digits tell the month? Why?
Which does tell the month?
Why do you think 86B21 means February?
Which pair of digits is most likely to tell the year? Why?

Write these dates in this code:

1 23rd April 1979   **3** 26th June 1989   **5** 10th December 1980
2 14th May 1981   **4** 15th March 1988   **6** 12th August 1987

34

6 length, journeys

EXAMPLE
100 cm = 1 metre   630 cm = 6.3 m
1000 m = 1 km   0.8 km = 800 m

A Give the answers in km.

1. 450 m + 1350 m
2. 1450 m + 590 m
3. 670 m + 890 m
4. 4400 m + 6600 m
5. 156 km − 34 km
6. 1040 km − 107 km
7. 3.5 m × 400
8. 5.6 m × 3000
9. 12.5 m × 200
10. 68 m × 400
11. 34.6 m × 40
12. 74 m × 20
13. 540 km ÷ 4
14. 608 km ÷ 20
15. 6000 km ÷ 500

B Give the answers in m.

1. 60 m + 200 cm
2. 45 m + 350 cm
3. 2.1 m + 600 cm
4. 27 m + 300 cm
5. 5.6 km − 5.4 km
6. 4.9 km − 4.85 km
7. 4 cm × 20
8. 6.2 cm × 400
9. 0.6 cm × 400
10. 0.3 cm × 60
11. 2.8 cm × 300
12. 34.5 cm × 50
13. 1 km ÷ 25
14. 20 km ÷ 4
15. 5 km ÷ 200

ROADS AND THEIR USAGE

C

	Length of road network (km)	Length of motorway (km)	Cars (thousands)	Lorries etc. (thousands)	Buses, coaches (units)
Austria	107 132	1 047	2 361	193	9 184
Belgium	126 990	1 317	3 230	257	18 744
Cyprus	10 948		104	30	1 497
Denmark	69 661	526	1 358	235	7 785
Norway	83 371	71	1 337	166	14 152
U.K.	342 925	2 946	15 632	1 880	110 000

1. Check that this table of statistics shows that there were 193 000 lorries in Austria and 110 000 buses and coaches in the U.K.

2. Which country has the longest road network, and how long is it to the nearest thousand km?

3. Write a list of countries showing the length of each road network to the nearest thousand km. Put this list into order according to length.

4. Make a list of the countries in the order of the length of their motorways.

5. Correct this statement: Austria has about 7000 more buses and coaches than cars.

6. Correct this statement: Norway has 21 000 more cars than Denmark.

7. If every bus and coach in Austria could be taken onto their motorways at the same time and fairly evenly spaced, about how many would there be on each kilometre? How many buses/coaches per kilometre would there be in each of the other countries?

8. If the average length of a car is 5 m and all the cars could be placed bumper to bumper in their own countries, which country would have the longest line? What would be the length of the shortest line?

ON A MOTORWAY

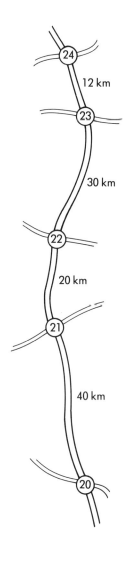

A On the Continent the distances between towns are measured in kilometres (km). On a motorway the junctions are numbered.

1 What is the shortest distance between two junctions shown on this stretch of motorway?

2 What is the longest distance between two junctions?

B On a motorway a car can usually travel at a steady speed. A car which is travelling at 100 km/h goes 50 km in half an hour. In $\frac{3}{4}$ hr it goes 75 km.

1 A car is travelling at 60 km/h. How far does it go in half an hour? How far does it go in 20 minutes? How far does it go in 2 hours?

2 A car is travelling at 120 km/h. How long will it take to go 60 km? How long to go 6 km? 10 km? 40 km?

3 An old car travels from junction 22 to 23 in half an hour. How far does it go? How fast is it going?

4 A red car travels from junction 22 to 23 in $\frac{1}{4}$ hour. How fast is it going?

5 A police car chases a stolen car from junction 23 to 21 in 20 min. How fast is it going?

6 A blue car goes from junction 23 to 24 in 10 minutes.
A grey car goes from junction 21 to 22 in $\frac{1}{4}$ hour.
A black van goes from junction 21 to 23 in $\frac{1}{2}$ hour.
How fast is the fastest vehicle going? Which is going slowest?

7 At 2.00 pm a driver passes junction 20. He wants to reach junction 21 by 2.30 pm. How fast must he travel?

8 A driver stops at a service station. He buys £15 worth of petrol. He takes his wife and son to the cafeteria. The boy's meal costs £2.15 and the grown-ups' meals are £4 each. How much does he spend?

9 Between junction 20 and 24 the speed limit is 90 km/h. Copy and complete the table below. Which of these cars are breaking the speed limit?

CAR	JOURNEY	DISTANCE	TIME	SPEED
green	junction 21 to 20		$\frac{1}{2}$ hour	
mauve	junction 24 to 23		6 minutes	
red	junction 23 to 21		$\frac{1}{2}$ hour	
yellow	junction 21 to 22		$\frac{1}{4}$ hour	

JOURNEYS

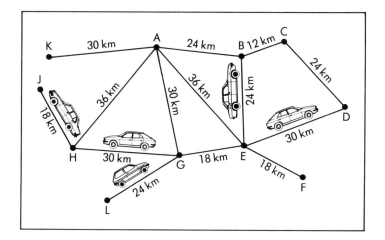

A

Litres	1	2	3	4	5	6	7	8	9	10	11	12
Km	12	24						96				144

This table shows that a car travels 12 kilometres on 1 litre of petrol. Copy and complete both the tables below.

B

Journey	Distance	Litres
A ►B	24 km	2
A ►E	36 km	3
1 D►F		
2 B►C►D		
3 A►B►C		
4 A►B►C►D		
5 B►A►G		

	Journey	Distance	Litres
6	K ►A ►B		
7	A►E►G		
8	G►H►J		
9	F►E►A►K		
10	C►B►E►A		
11	K►A►G►L		
12	F►E►G►L		

C All these journeys are through a busy town. It is estimated that the average speed is about 30 km/h.

How long should be allowed for these journeys?

1 A to K
2 A to B
3 A to H
4 A to G to E
5 A to D via B and C

6 H to L via G
7 K to G via A
8 H to B via A
9 G to D
10 F to B via E

ESTIMATING FOR A JOURNEY

A **1** It took a salesman 2 hours 55 minutes to travel 180 miles. Estimate his average speed.

2 A car used about 30 litres of petrol on a 200 mile journey. The petrol cost 44.8p a litre. About how much would the journey have cost?
How many miles to the litre was the car doing?

3 Mr Jones started his journey at 9.15 am and arrived at 11.45 am. He thought he had managed to average 50 mph. Which is the most likely distance he had travelled: 100 miles, 130 miles or 150 miles?

4 The pilot of an aircraft tells his passengers that the flight is expected to take $2\frac{1}{4}$ hours and the average speed is 450 mph. How far will they be flying?

5 A lorry driver started his journey at 7.35 am. When he stopped at 9.30 am he had travelled 220 km.
Estimate his average speed.
Was it 110 km/h, 130 km/h or 140 km/h?
He had a 15 minute break before resuming his journey. He was then able to travel at a steady speed of 115 km/h for $1\frac{1}{2}$ hours to his destination. About how many kilometres was this?
When did he arrive at his destination?

6 The petrol tank on my car holds 12 gallons when full. The car does 31 miles to the gallon. The petrol gauge shows the tank is $\frac{1}{4}$ full. How much petrol is left?
How far, to the nearest 10 miles, should I be able to travel?
How much petrol would it use to travel 120 miles, to the nearest gallon?

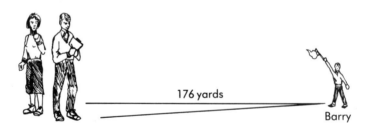

176 yards

Barry

B A group of children did a survey of the speed of cars passing their school. Barry stood 176 yards (176 yards = $\frac{1}{10}$ mile) along the straight stretch of the road away from the other children.
As each car passed Barry they timed how long it took to travel 176 yards. Here are some of their observations:
red car 10 seconds; green van 15 seconds; ambulance 9 seconds.
In the classroom they worked out the speeds. What were they?
They realised that they wanted to know the speed while they could still see the car, so they worked out a table.

They used a formula:
360 ÷ no. of seconds = speed in mph.

Time to travel 176 yards	Speed
15 seconds	24 mph
12 seconds	
10 seconds	
9 seconds	
8 seconds	
6 seconds	

Copy and complete the table.

FAST AND SLOW

A 1 Kathy decided to walk to her friend's house 1½ miles away. She left home at 6.30 pm and arrived at 7.00 pm. How fast did she walk?

2 A man has been driving along the motorway at a steady speed of 70 mph for 1½ hours. How far has he travelled?

3 The flight between two cities took 45 minutes. The pilot said they had been travelling at an average speed of 320 mph. How far apart were the two cities?

4 A long distance lorry driver has already done 70 miles of his 350 mile journey. How much further has he to travel? He expects to be able to drive at a steady speed of 60 mph for the rest of the journey. He will have to stop for a half hour rest. How much longer will it be before he reaches his destination?

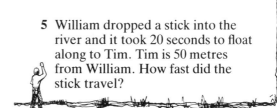

5 William dropped a stick into the river and it took 20 seconds to float along to Tim. Tim is 50 metres from William. How fast did the stick travel?

6 A power drill was being used to cut through a piece of metal 20 mm thick. It took exactly 10 seconds. How fast did the drill penetrate the metal?

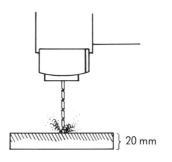

} 20 mm

AN OLD BUS TIMETABLE

B This chart shows that the bus left Dewsbury at 9.40 am when it was going South.
R meant that there was a stop for refreshments.

1 On the southbound journey, when did it stop for refreshments? At that time of day would the stop have been for lunch, tea or supper?

2 Where did it stop for refreshments when northbound?

3 When did it leave Halifax? How long did the journey between Halifax and Huddersfield take?

4 How long did the journey from Birmingham to Halifax take?

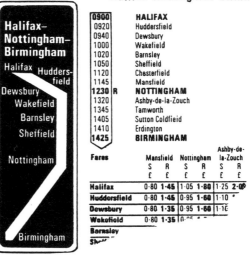

Halifax–Nottingham–Birmingham

0900	HALIFAX	2035
0920	Huddersfield	2015
0940	Dewsbury	1955
1000	Wakefield	1935
1020	Barnsley	1915
1050	Sheffield	1845
1120	Chesterfield	1815
1145	Mansfield	1750
1230 R	NOTTINGHAM	1715 R
1320	Ashby-de-la-Zouch	1615
1345	Tamworth	1550
1405	Sutton Coldfield	1530
1410	Erdington	1525
1425	BIRMINGHAM	1510

Fares	Mansfield		Nottingham		Ashby-de-la-Zouch	
	S £	R £	S £	R £	S £	R £
Halifax	0·80	1·45	1·05	1·80	1·25	2·0
Huddersfield	0·80	1·45	0·95	1·60	1·10	
Dewsbury	0·80	1·35	0·95	1·60	1·1	
Wakefield			0·80	1·35		
Barnsley						
Sh						

5 The table shows that the single fare from Dewsbury to Nottingham was 95p. How much was the return fare between Dewsbury and Nottingham?

6 How much cheaper was it to buy a return ticket than to buy single tickets for the journey from Huddersfield to Mansfield and back?

GRAPHS OF A JOURNEY

A   A hiker leaves A at 9 am and walks for
one hour.

1 How far from A is he by 10 am?
2 How fast did he walk?
3 At 10 am he rests. How long does he rest?
4 How long does he take to walk the next
kilometre?
5 When does he stop again?
6 How far is he then from home?
7 If he had taken 3 hours to do the journey,
and had not stopped for any rests, how
fast would he have had to walk?

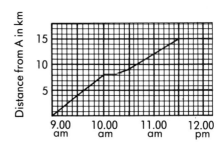

B   1 What time does the bus leave village A?
2 It goes 10 km to village B. When does it arrive?
3 How long does it stay at B?
4 How far is it to the next stop?
5 How fast did he travel?
6 How long was the driver's break before
starting the return journey?
7 How fast did he travel to B?
8 When did he get back to A?
9 How fast did he travel from B to A?

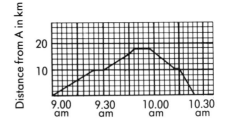

C   Two people set out from places 12 km
apart and walk to meet each other.
1 Who reached the meeting place first?
2 How far was the meeting place from A?
3 How fast did each person walk?
4 How long were they together?
5 How fast did they walk on their return
journey?

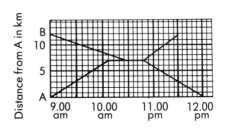

D   The two cars X and Y set off together.
1 How far did X travel before stopping?
2 How fast did X travel before stopping?
3 When did Y pass X?
4 When X restarted, he travelled fast and
overtook Y. When did he overtake Y?
5 How fast was X travelling?
6 How fast was Y travelling?
7 Who reached the finishing point first?
8 How far was the finishing point from the start?

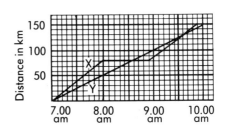

SCALE DRAWING

A Measure these lines.
Calculate the length they represent.

1 1 cm represents 5 metres

2 1 cm represents 3 kilometres

3 1 cm represents 10 km

4 1 cm represents 2 miles

5 1 cm represents 2 feet

B Calculate the lengths represented by these lines

	Length of line on map	Scale	Length represent- ed
1	2 cm	1 cm rep. 10 km	
2	3 cm	1 cm rep. 25 km	
3	3.2 cm	1 cm rep. 4 m	
4	5.1 cm	1 cm rep. 10 m	
5	8.7 cm	1 cm rep. 20 km	

C

Scale: 1 cm represents 10 ft

Mr Holmes has drawn this plan of his house and garden.

1 How wide is his piece of land?
2 How long is the site?
3 What is the total area of Mr Holmes' land?
4 How far is the front of the house from the front boundary?
5 What is the area of land occupied by the house and garage?
  What fraction of the site is occupied by the building?
6 What percentage of the site is used for growing vegetables?

7 What is the area of the lawn?
  What percentage of the site does the lawn occupy?
8 Mrs Holmes is trimming the edge of the lawn. In the first 5 minutes she has done 13 feet. How much longer will it take her to trim all round the lawn?
9 When will the sun be on the front of the house?
10 Why do you think they have built the patio against the wall?
11 Make a sketch of the floor plan of the house and show on it the measurements of each room.

SIGHTSEEING

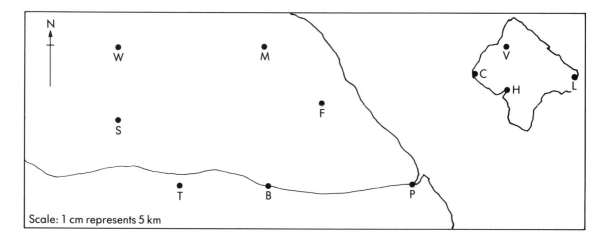

Scale: 1 cm represents 5 km

This part of a map shows a stretch of coastline and a small island. On the island there is a village V and a harbour H.

1 How far is V from H?
What is the direction of V from H?

2 There is a coastguard station at C and a lighthouse at L. How far apart are they?

3 How far is it between the most northerly and most southerly points of the island?

4 If this rectangle just encloses the island, mark the lengths of its sides in km.
What is the area of this rectangle? Make a statement about the area of the island.

5 A boat sets out towards P from the harbour H. In what direction will it be heading?

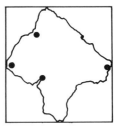

A helicopter is based at P. It has been hired to take photographs from the air of some of the landmarks of the region.

6 In what direction must the helicopter travel from P to the fort F?

7 From the fort F it heads towards the monument M. How far is it from F?

8 At W there is an old waterwheel. People waiting at W are looking for the helicopter. In which direction should they be looking?

9 From the waterwheel the helicopter travels SE. What is the destination?

10 How far is the station S from the old bridge at B?

11 When the helicopter leaves the station S it is on a bearing 105°. Where is it going?

12 A separate journey is planned from P to the tower at T. How long is this return journey?
It usually takes the helicopter 20 minutes to make the journey from P to T and back to P. What is the average speed?

7 weight, clocks and dials

THE SAME BUT DIFFERENT

1 kilogram = 2.2 pounds

A Use this approximate equivalent to make these conversions:

1 5 kg to pounds
2 7 kg to pounds
3 9 kg to nearest pound
4 12 kg to nearest pound
5 60 kg to nearest pound
6 50 kg to nearest pound
7 10 lb to nearest half kg
8 25 lb to nearest half kg
9 70 lb to nearest kg
10 100 lb to nearest kg

B Use <, > or = to make true statements

1 5 kg × 2   20 pounds
2 10 kg × 3   65 pounds
3 1.5 kg × 6   20 pounds
4 0.4 kg × 5   5 pounds
5 1.2 kg × 2   5 pounds
6 4 lb × 7   12 kg
7 15 lb ÷ 3   3 kg
8 2 lb × 8   8 kg
9 15 lb × 2   15 kg
10 16 lb ÷ 4   5 kg

C These four closed boxes look the same, but when they are put on the scales, this happens:

Which is heavier, A or B?
When B and C are compared, this happens:

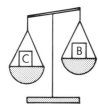

Complete the sentence about B and C:
So . . . is heavier than . . .
Which is lightest, A, B or C? Why?
What else would you need to do to find the heaviest?

The other box, D, also looks the same as A, B and C, but the balance shows

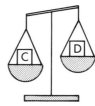

What else would you have to do to put D in order of weight with the others?

THE SAME PROPORTIONS

1

Raspberry jam

4 lb caster sugar
3½ lb fruit

sugar 2 kg

sugar 2 kg

sugar 2 kg

I have picked 7 lb of fruit. I know that 1 kg is about 2.2 pounds. How many of these bags will I need?

2

Cake recipe

3 eggs
150 g flour
150 g sugar
150 g margarine

best white flour 1.5 kg

sugar 1 kg

The teacher is going to buy enough flour and sugar for each pupil in her cookery class to make a cake. There are 12 pupils in the class. How many of these bags of flour and sugar should be ordered? (1 kg = 1000 g)

3

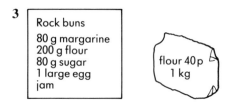

Rock buns

80 g margarine
200 g flour
80 g sugar
1 large egg
jam

flour 40 p 1 kg

This recipe makes 12 buns. Write out the recipe for 36 buns.
How much would the flour for 12 rock cakes cost? (1 kg = 1000 g)

4 A survey showed that the average consumption of crisps in the U.K. was 2.1 kg per annum per person. How much is this per week?
Crisps are sold in 40 g packets. How many packets is this weekly average?
The same survey showed that on average each person in the U.K. eats 7.3 kg of chocolate in a year. How many 50 g bars is this?

5 Kathy wants to try her grandma's recipe for a fruit loaf:

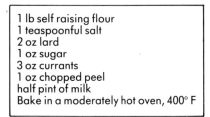

1 lb self raising flour
1 teaspoonful salt
2 oz lard
1 oz sugar
3 oz currants
1 oz chopped peel
half pint of milk
Bake in a moderately hot oven, 400° F

Rewrite the recipe for Kathy so that she can make one half the size of her grandma's. (1 lb = 16 oz)

CLOCKS AND DIALS

1  The chart on this fryer is

	minutes	°C
mushrooms	3	150
fritters	5	160
fish	5	170
chicken	15	180
chips	12	190

What are these dials set for cooking?

| 150 | 160 | 170 | 180 | 190 |

2 These time controllers have been set to switch on and off. They also tell the time of day to the nearest half hour.
What time is each clock showing?

When will each switch on and off in the morning? When will they come on in the evening, and how long will they be on?

 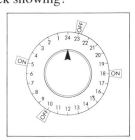 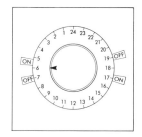 

3 The maker of these scales claims that they are accurate to within 1% of the total weight recorded. What is the weight recorded on the dial? What is the least true weight it could be?

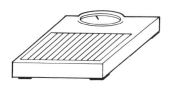 

4 What were these watches showing ½ minute ago?
What will they be showing in 15 seconds from now?

(i)  (ii)  (iii)  (iv)

5 Write these in everyday language: The plane is due to land at 17.45 hours.
The ship sailed at 1500 hours. The train departs at 16.30 hours.

CLOCKWISE AND ANTICLOCKWISE

A This dial shows 3 and a bit:

1 per division

and this shows just over 20:

Together these dials show 23 and a small bit more.

10 per division

Choose the number which best describes what each pair of dials is showing:
Remember that the dials are recording a number which is increasing, so write down the number which the pointer has passed.

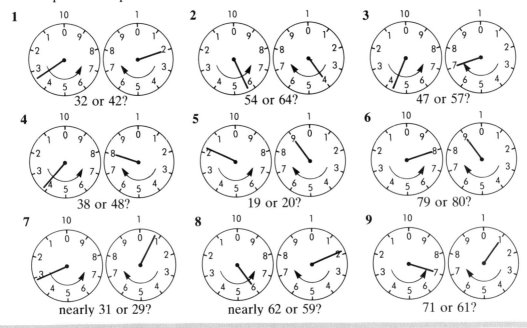

1 32 or 42?

2 54 or 64?

3 47 or 57?

4 38 or 48?

5 19 or 20?

6 79 or 80?

7 nearly 31 or 29?

8 nearly 62 or 59?

9 71 or 61?

B This meter has the dials arranged with some pointers going clockwise, but others turn anticlockwise.

1 This meter reading was given as 51263 but one of the dials has been read incorrectly. What should the reading be?

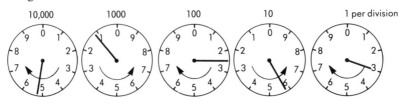

2 What reading do these dials give?

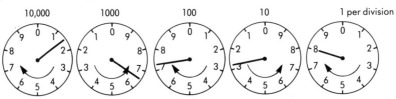

46

ELECTRICITY

A Electricity is measured in units called kilowatt hours, kWh.

If a unit costs about 6p, calculate the approximate cost of:

ONE UNIT PROVIDES approximately	1 kW fan heater 1 hour	2 bar electric fire ½ hour	colour TV 9 hours
	stereo system 8 hours	tumble dryer ½ hour	shower 10 minutes

1 running a fan heater for 3 hours.
2 running the tumble dryer for 1 hour.
3 watching TV for about 4 hours every day for one week.
4 playing the stereo for 4 hours on Saturday.
5 having a 5 minute shower daily for 1 week.
6 burning a 2 bar electric fire for 5 hours each day for a week.
7 running a fan heater for 1 hour daily for 1 week.
8 using the tumble dryer for 2 hours.
9 listening to the stereo for 24 hours.
10 having a 5 minute shower on 4 days.

B 1 Many householders choose to pay for their electricity quarterly. Every three months they will receive a calculated account. Basic quarterly charge of £8.60 plus 5.77p for each unit used. Copy and complete these accounts:

Mr Jones	
Basic charge	£8.60
1000 units @ 5.77p	£
Total	£

Mr Brown	
Basic charge	£
1543 units @ 5.77 p	£
Total	£

2 Some people prefer to pay monthly. The amount to be paid is calculated by using the total for the previous year and assuming the same number of units will be used again. The monthly payment is made to the nearest pound. At the end of the year there may be a refund or extra payment needed.
Last year Mr Smith's quarterly accounts were £63.37, £40.25, £39.60 and £56.20. What was his total for the year?

He decides to make monthly payments for the coming year. How much will he pay?

Mr Smith buys a dishwasher which uses 20 units weekly.
He has the dishwasher for the last 10 weeks of the year.
How much extra should he be prepared to pay for the electricity it has used?

C There is a cheaper tariff available when more than 10% of the units can be used during the night. It means having two meters and a higher basic charge.

ECONOMY TARIFF	
Basic quarterly charge	£10.85
Daytime units	6.10 p
Night-time units	2.07 p

These are Mr White's meters for a quarter. How many units have been used on each meter? Copy and complete his account.

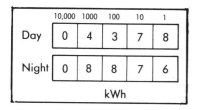

	10,000	1000	100	10	1
Day	0	4	3	7	8
Night	0	8	8	7	6

kWh

	10,000	1000	100	10	1
Day	0	5	1	7	8
Night	0	9	0	2	6

kWh

Mr White	
Basic charge	£
units @ 6.10 p	£
units @ 2.07 p	£
Total	£

8 area

AREAS ON A GRID

Which of these triangles have the same area as triangle P?

1

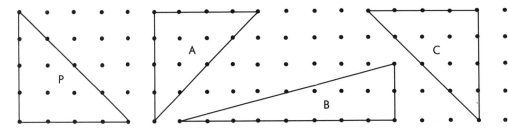

Which of these triangles have the same area as shape Q?

2

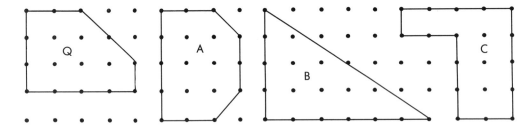

Which of these shapes have the same area as shape R?

3

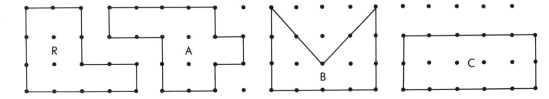

Which of these shapes have the same area as shape S?

4

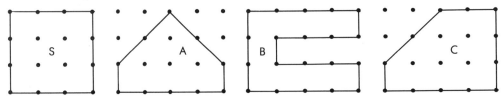

Which of these shapes have the same area as triangle T?

5

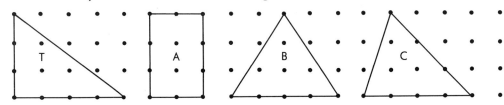

PAPER SLEEVES

1

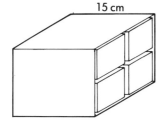

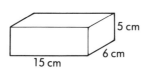

A manufacturer wraps four packets together in a paper sleeve.
1 cm overlap is allowed for sticking it together.
Which of these drawings would make a sleeve which fits?

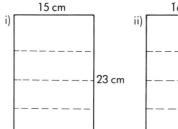
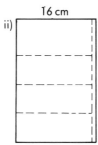
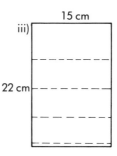

Calculate the area of paper used to make one sleeve.
How much paper would be needed to make 100 sleeves?

2

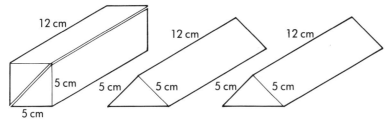

These packets are wrapped in pairs. Make a sketch of the shape of the paper needed to make the sleeve. Allow 1 cm overlap. Show on it the measurements.
Calculate the area of paper needed for one sleeve.

3

These packets are wrapped in sixes. An overlap of 1 cm is allowed. Sketch the paper needed for the sleeve and calculate the area of paper used for one wrapper.

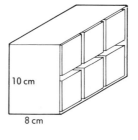
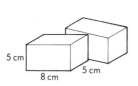

4

These cubes are wrapped in blocks of eight. Sketch the paper needed for a sleeve, with 1 cm overlap. Calculate the area of paper used for each wrapper.

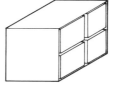

49

WALLS, CEILINGS AND FENCES

1 These diagrams show the doors and windows on each of the walls of a room. The walls have to be painted with emulsion paint. Calculate the area to be painted.

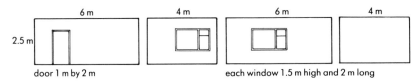

The emulsion paint is sold in 3 litre cans. Each can is said to cover 30 sq.m. How many cans will be needed to give these walls one coat?

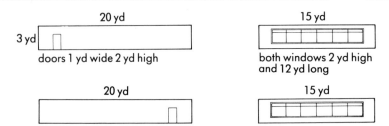

2 This diagram shows the walls of a hall which has to be painted.
Make a sketch of the ceiling of this hall and calculate its area.
Calculate the area of the wall surfaces which have to be painted.
The ceiling also has to be painted with the same paint. What is the total area to be painted?
The walls and ceiling need two coats. What is the number of square yards to be covered?
The paint is sold in tins costing £10 and each tin will cover about 80 square yards.
How many tins will be needed?
How much will it cost to repaint this hall?

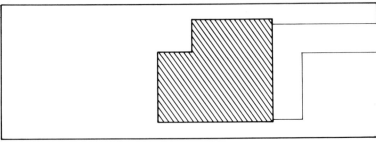

Scale: 1 cm represents 10 m

3 This is the plan of a house garden. The fence along the sides and back is 1 metre high. Both sides of it have to be treated with preservative which comes in cans sufficient to cover an area of 50 sq. metres. How many cans will be needed for this fence?

50

THE TRAPEZIUM

A

1

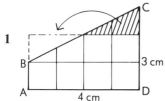

ABCD is a trapezium.
When the shaded triangle is moved, the shape is a rectangle.
The area of the trapezium is the same as the area of the rectangle.
What is the area of the rectangle?
What is the area of the trapezium?

Draw the rectangle which can be made by moving the shaded triangle on each trapezium:

2

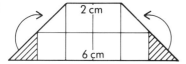

Average of parallel sides =
Distance apart of parallel sides =
Area of trapezium =

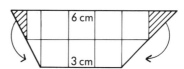

Average of parallel sides =
Distance apart of parallel sides =
Area of trapezium =

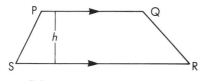

Draw this shape on squared paper.
Make sure the parallel sides are 6 cm and 10 cm long.
They are . . . cm apart.
A and B are the centres of the sloping sides.
How long is AB? AB is $\frac{1}{2}$ (. . . + . . .)

3 Copy these shapes. Measure the lines joining the centres of the sloping sides.
Check that it is always $\frac{1}{2}$ (. . . + . . .). Calculate the area of each trapezium:

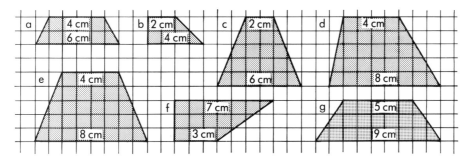

B

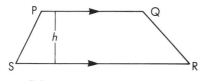

PQRS is a trapezium.
Use the formula area = $\frac{1}{2}(a + b)h$ to calculate the area
of these:

	PQ	RS	height		PQ	RS	height
1	3 cm	7 cm	4 cm	**6**	12 cm	14 cm	8 cm
2	5 cm	7 cm	5 cm	**7**	10 cm	8 cm	4.5 cm
3	9 cm	7 cm	5 cm	**8**	21 cm	19 cm	13.5 cm
4	8 cm	9 cm	6 cm	**9**	12 cm	15 cm	8 cm
5	8 cm	12 cm	7.8 cm	**10**	16 cm	12 cm	14.6 cm

51

SQUARES

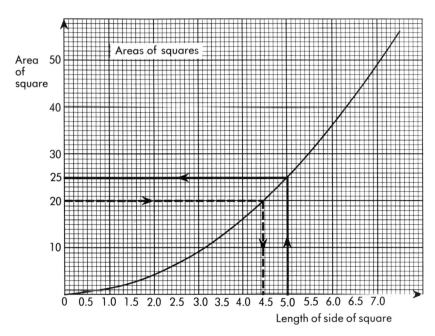

Area of square (vertical axis): 10, 20, 25, 30, 40, 50

Length of side of square (horizontal axis): 0, 0.5, 1.0, 1.5, 2.0, 2.5, 3.0, 3.5, 4.0, 4.5, 5.0, 5.5, 6.0, 6.5, 7.0

Areas of squares

A   The area of this square is $5 \times 5 = 25$ units.
Find the line on the graph which shows side of square 5 cm.
Follow the arrow up to the curve and then across the page to 25.
This curve works out areas.
Check that it shows $4 \times 4 = 16$.

1 Use it to find 5.5×5.5. What is the area of a square of side 5.5 cm?
2 Use it to find 2.5×2.5. What is the area of a square of side 2.5 cm?
3 Use it to find 5.9×5.9. What is the area of a square of side 5.9 cm?
4 Use it to find 4.2×4.2. What is the area of a square of side 4.2 cm?
5 Use it to find 3.5×3.5. What is the area of a square of side 3.5 cm?

B A square has an area of 20 cm^2. Follow the dotted arrow. It shows its side must be 4.4 cm.

1 The area of a square is 10 cm^2. How long is its side?
2 The area of a square is 30 cm^2. How long is its side?
3 The area of a square is 36 cm^2. How long is its side?

C A computer has been used to draw a square on the screen.
The sides of the square are made to grow.

1 Will the area increase or decrease?
2 When the side of the square grows from 3 cm to 4 cm, how much has the area of the square changed?
3 How much does the area change when the side grows from 4 cm to 5 cm?
4 How much did the area change when the side grew from 3 to 3.5 cm?
5 Describe the way in which the area of the square grows.

CIRCLES

A

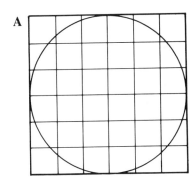

If each square of the grid represents 1 cm², the radius of this circle is 3 cm.
Count the squares to check it covers about 28.
Then its area is 28 cm².
Now find the radius 3 cm on the graph and follow the arrows. The graph shows its area is 28 cm².

1 Find 2.5 cm on the graph and follow the arrow up to the graph. What does it say the area is?
2 If the radius is 3.5 cm what does the graph give for the area of the circle?
3 The radius of a circle is 1.5 cm. What is its area?
4 The radius of a circle is 2.1 cm. What is its area?
The area of a circle is 40 cm². Check that the graph says its radius must be about 3.6 cm.

5 The area of a circle is 45 cm². What does the graph say its radius must be?
6 The area of a circle is 30 cm². What does the graph say its radius must be?
7 The area of a circle is 10 cm². What does the graph say its radius must be?

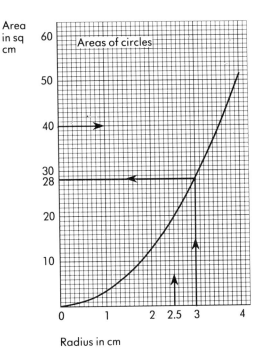

Look at the square which has been drawn round the circle at the top of the page.
How long is each side of the square?
How many squares does it cover?
What is the area of that square?
The area of the square is . . . than the area of the circle.

8 A circle has radius 2 cm.
A square is drawn round it.
What is the area of this square?
9 A circle has radius 5 cm.
A square is drawn round it.
What is the area of this square?

B Use the formula area of circle = πr^2 ($\pi = 3.1$) to find the area of:
1 a circle, radius 3 cm
2 a circle, radius 10 cm
3 a circle, radius 9 cm
4 a circle, diameter 8.4 cm
5 a circle, diameter 15 cm
6 a semi circle, radius 5 cm
7 a semi circle, radius 10 cm
8 a semi circle, radius 7.5 cm
9 a semi circle, diameter 5.2 cm
10 a semi circle, diameter 18.6 cm

53

A VARIETY OF SHAPES

Area of circle = πr^2 Area of parallelogram = base × perpendicular height
$$= b \times h$$

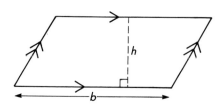

Area of triangle = $\frac{1}{2} b \times h$

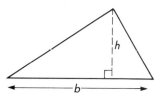

Calculate the area of each shape. Use $\pi = 3.1$

A

1

10 cm

2

5 cm

3

12 cm

4
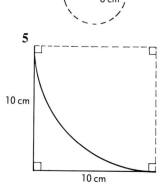
8 cm

5
10 cm

10 cm

B

1
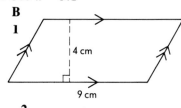
4 cm
9 cm

2
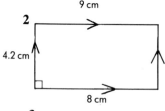
4.2 cm
8 cm

3
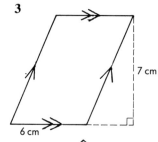
7 cm
6 cm

4
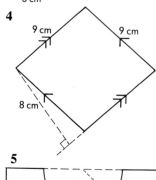
9 cm   9 cm
8 cm

5
6 cm
8 cm
12 cm

C

1
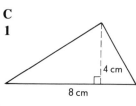
4 cm
8 cm

2
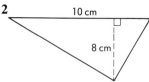
10 cm
8 cm

3
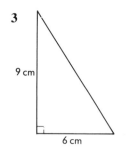
9 cm
6 cm

4
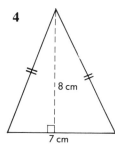
8 cm
7 cm

5
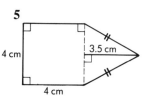
4 cm   3.5 cm
4 cm

54

PERIMETER AND AREA

1 Measure the perimeter of each figure.   **2** Find the area of each figure. Tabulate the results.

What could you say about the area of a figure with 10 equal sides and the same perimeter as these?

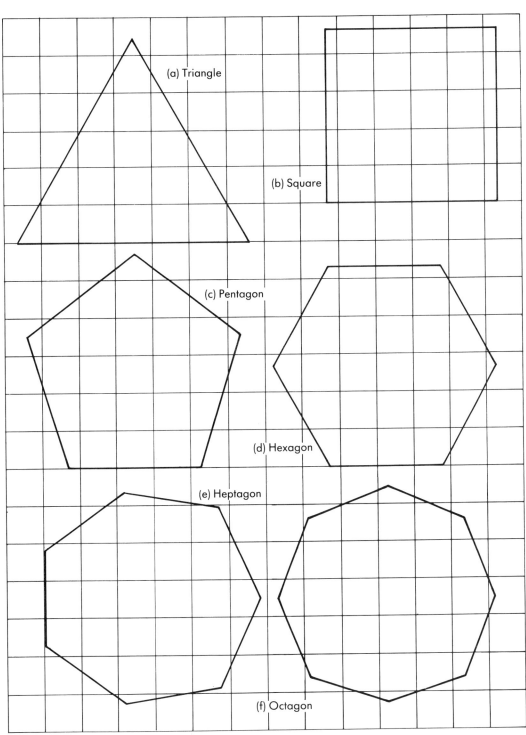

A PATTERN FROM AREA

Use centimetre squared paper when a copy of any of these diagrams is needed.

EXAMPLE

The lines of **a** have been increased by a factor of 2
because each line in shape **b** is twice as long as in **a**.
The area of **a** is $3\,\text{cm}^2$.
When **a** was enlarged to **b** its area was increased by a factor of 4

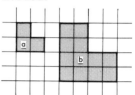

A Make an enlargement of **c**, **d** and **e** so that each length
is increased by a factor of 2.
Find the area of each enlargement.

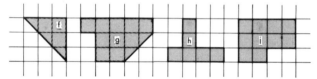

Copy and complete this table:

Shape	Length enlargement factor	Area of original shape	enlargement	Area enlargement factor
a	2	3	12	4
c	2	2		
d	2			
e				

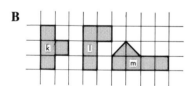

What is the area of these shapes?
If the sides were doubled in length,
what would the areas become?
Enter the results in the table.

B

Make enlargements of these shapes so that
each length is increased by a factor of 3.
Find the area of each enlargement.
Copy and complete this table:

Shape	Length enlargement factor	Area of original shape	enlargement	Area enlargement factor
k	3			
l				
m				

C Enlarge this shape so that each length becomes four times as long.
What is the area of the new shape?
What would its area have been if the length of the
sides had been increased 5 times?

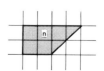

9 volume

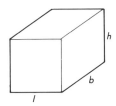

Volume of cuboid =
length × breadth × height

Volume of cylinder = $\pi r^2 h$

A Find the volume of the following rectangular solids:

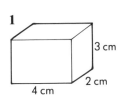

1 3 cm, 2 cm, 4 cm

2 4 cm, 2 cm, 1 cm

3 3 cm, 3 cm, 5 cm

4 1.5 cm, 4 cm, 1.5 cm

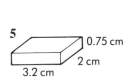

5 0.75 cm, 2 cm, 3.2 cm

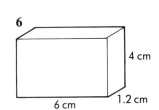

6 4 cm, 1.2 cm, 6 cm

7 4 cm, 4.2 cm, 2.1 cm

8 6 cm, 4.2 cm, 3.2 cm

B Calculate the volume of these cylinders. Use $\pi = 3.1$

1 $r = 5\,\text{cm}, h = 12\,\text{cm}$	**6** $r = 3.5\,\text{cm}, h = 12\,\text{cm}$	**11** $r = 1\,\text{m}, h = 2\,\text{m}$
2 $r = 6\,\text{cm}, h = 8\,\text{cm}$	**7** $r = 2.1\,\text{cm}, h = 5\,\text{cm}$	**12** $r = 1.5\,\text{m}, h = 4\,\text{m}$
3 $r = 4\,\text{cm}, h = 10\,\text{cm}$	**8** $r = 0.7\,\text{cm}, h = 12\,\text{cm}$	**13** $r = 0.5\,\text{m}, h = 10\,\text{m}$
4 $r = 3\,\text{cm}, h = 11\,\text{cm}$	**9** $r = 0.6\,\text{cm}, h = 15\,\text{cm}$	**14** $r = 0.1\,\text{m}, h = 20\,\text{m}$
5 $r = 2\,\text{cm}, h = 8\,\text{cm}$	**10** $r = 0.8\,\text{cm}, h = 10\,\text{cm}$	**15** $r = 0.2\,\text{m}, h = 25\,\text{m}$

Suggest cylindrical objects for some of these measurements.

C

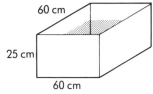

This tank contains water to a depth of 15 cm.
1 How much water is in the tank?
2 How much more water could be poured in before it overflows?
3 Each edge of a heavy cube is 30 cm long.
 The cube is placed gently into the water.
 How much will the water level rise? Will it overflow?

VOLUME

For a solid with a uniform cross-section, Volume = area of cross-section × length

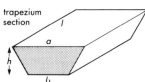

trapezium section

Area of cross-section = $\frac{1}{2}(a + b) \times h$

EXAMPLE

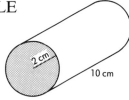

Area of cross-section = $3.1 \times 2 \times 2 \, \text{cm}^2$
$$= 3.1 \times 4 \, \text{cm}^2$$
$$= 12.4 \, \text{cm}^2$$
Volume = $12.4 \times 10 \, \text{cm}^3$
$$= 124 \, \text{cm}^3$$

EXAMPLE

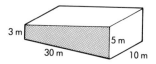

Area of cross-section = $\frac{1}{2}(3 + 5) \times 30 \, \text{m}^2$
$$= 4 \times 30 \, \text{m}^2$$
$$= 120 \, \text{m}^2$$
Volume = $120 \times 10 \, \text{m}^3$
$$= 1200 \, \text{m}^3$$

A Calculate the volume of these cylinders:
Use $\pi = 3.14$

cross-section	length
1 $12 \, \text{cm}^2$	7 cm
2 $35 \, \text{cm}^2$	8 cm
3 $r = 6 \, \text{cm}$	10 cm
4 $r = 1 \, \text{cm}$	6 cm
5 $5 \, \text{cm}^2$	6 cm
6 $r = 3 \, \text{cm}$	10 cm
7 $r = 2 \, \text{cm}$	6 cm
8 $r = 10 \, \text{cm}$	20 cm
9 $5 \, \text{cm}^2$	5.5 cm
10 $4 \, \text{cm}^2$	12 cm

B Calculate the volume of these troughs:

	parallel sides	distance apart	length
1	3 m, 7 m	12 m	15 m
2	4 m, 8 m	3 m	20 m
3	8 m, 12 m	6 m	100 m
4	5 cm, 9 cm	10 cm	8 cm
5	18 cm, 22 cm	30 cm	50 cm
6	24 m, 26 m	2 m	50 m
7	6 cm, 16 cm	8 cm	25 cm
8	2 m, 1 m	60 m	20 m
9	1 cm, 9 cm	5 cm	4 cm
10	25 m, 35 m	2 m	100 m

C Cans, each 6 cm diameter and 15 cm tall have to be packed into cartons with the measurements shown. How many cans can be fitted into one carton? How many cartons are needed for 1200 cans?

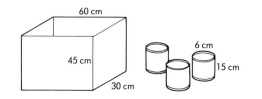

LIQUIDS

1 gallon = 4.5 litres, approximately
8 pints = 1 gallon

100 centilitres = 1 litre, 100 cl = 1 l
1000 millilitres = 1 litre, 1000 ml = 1 l

A One of these size labels belongs on each container.
Choose the most likely one for each.

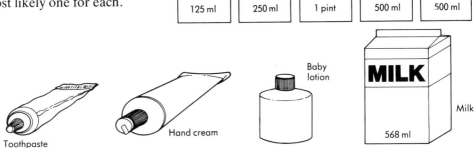

| 125 ml | 250 ml | 1 pint | 500 ml | 500 ml |

Cough mixture

Toothpaste

Hand cream

Baby lotion

MILK

Milk

568 ml

B Use the equivalents given above to express:

1 9 litres in gallons
2 45 litres in gallons
3 100 litres in gallons
4 80 litres in gallons
5 72 litres in gallons
6 5 gallons in litres
7 13 gallons in litres
8 6 gallons to litres
9 12 gallons to litres
10 50 gallons to litres

11 500 ml in litres
12 1500 ml in litres
13 250 ml in litres
14 $\frac{1}{2}$ litre in ml
15 $1\frac{1}{2}$ litres in ml
16 $\frac{1}{2}$ gallon in pints
17 $1\frac{1}{2}$ gallon in pints
18 4 pints to litres
19 1 pint to litres
20 1 pint in ml

C

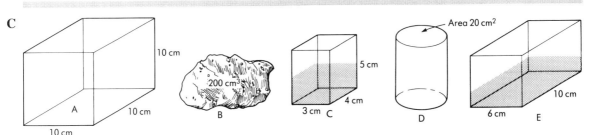

1 Calculate the volume of the tank A.
   $1000 \, cm^3 = 1$ litre. Give another name for cm^3.
   How many ml of water will the tank A hold?
2 The tank A is full of water and the stone B is gently lowered into the water.
   How much water will overflow?
3 How much water is in the tank C?
   All the water from tank C is poured into cylinder D. The area of the base of D is $20 \, cm^2$.
   What will be the depth of the water in D?
4 The tank E has some water in it. The cube F is lowered carefully into the water and is
   covered by the water. What is the volume of the cube?
   The area of the base of tank E is $60 \, cm^2$. How much will the cube make the water level rise?

59

10 geometry

BEARINGS

EXAMPLE

Three figure bearings

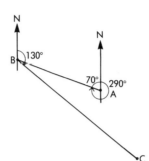

Looking from O, P is in the direction 045°
P is in the direction N E

Looking from A in direction 290°, B is in sight
From B, the direction towards C is 130°

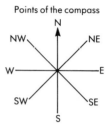

Points of the compass

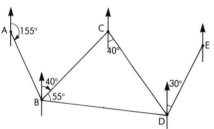

A Complete using 3 figure bearings:
1 From A, B is in the direction
2 From B, C is in the direction
3 From B, D is in the direction
4 From D, C is in the direction
5 From D, E is in the direction
6 From E, D is in the direction
7 From D, B is in the direction
8 From C, B is in the direction
9 From C, D is in the direction
10 From E, A is in the direction

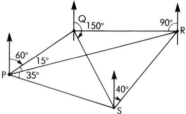

B Complete using 3 figure bearings:
1 From P, Q is in the direction
2 From P, R is in the direction
3 From P, S is in the direction
4 From Q, P is in the direction
5 From Q, R is in the direction
6 From Q, S is in the direction
7 From S, R is in the direction
8 From R, Q is in the direction
9 From S, P is in the direction
10 From S, Q is in the direction

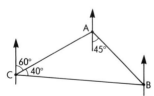

C Use the points of the compass for
1 the direction of A from C
2 the direction of C from A
3 the direction of A from B
4 the direction of B from A
5 the direction of B from C

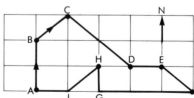

D The diagram shows a cross country path.
1 Which point is furthest North?
2 Which point is furthest East?
3 From A go N to B.
   What is the direction from B to C?
4 Make a list of the directions in which to travel
   round the rest of the path and back to A.

MEASURING ANGLES

A Use a protractor to make an accurate measurement of each marked angle.

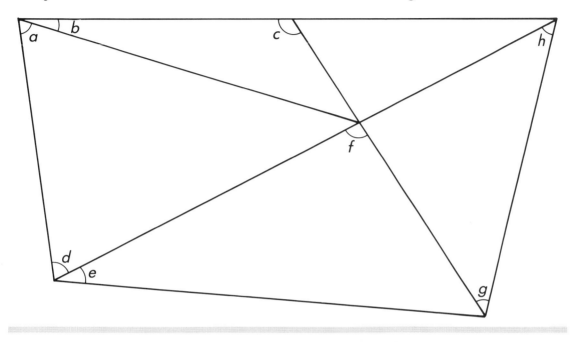

B Make an accurate measurement of these angles. Say if the angle is obtuse.

1 ∠BAK   **3** ∠EDF   **5** ∠KFE   **7** ∠ABK
2 ∠BKC   **4** ∠CBK   **6** ∠AEC   **8** ∠ACE

THE LANGUAGE OF SHAPE

Give (i) the name of each shape and say whether it has any symmetry and what sort.
    (ii) the marked measurement or part.
   (iii) the position of the solid figures, horizontal or vertical.

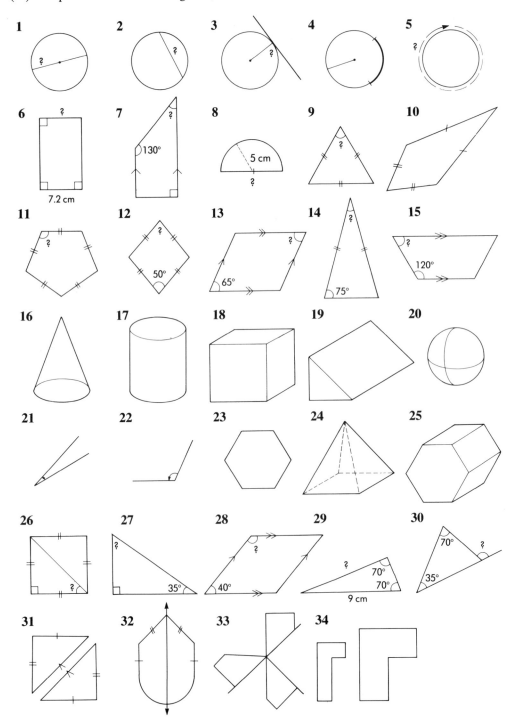

ANGLES AND PARALLELS

EXAMPLE

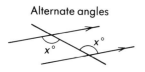

Alternate angles

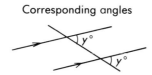

Corresponding angles

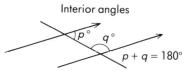

Interior angles

$p + q = 180°$

Calculate the angles marked with a letter.

1

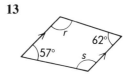

2

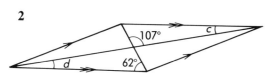

3

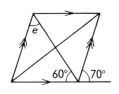

4

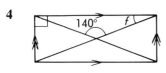

5

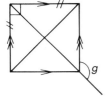

6

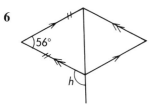

7

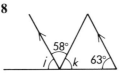

8

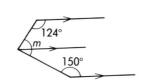

9

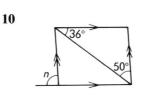

10

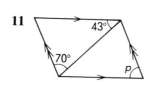

11

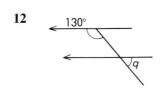

12

13

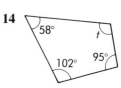

14

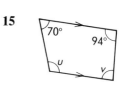

15

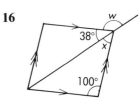

16

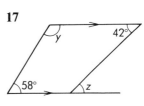

17

18

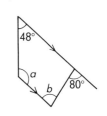

TRANSFORMATIONS

A Describe each pair of shapes, using translation, rotation, reflection or enlargement.

i)

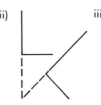

ii)

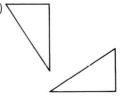

iii)

iv)

v)

vi)

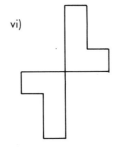

B Copy each shape on to squared paper and reflect it in the mirror line.

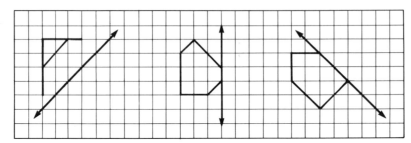

C

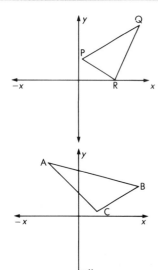

1 Draw triangle PQR on squared paper. The coordinates are
P (1,4)   Q (10,10)   R (6,0)
Reflect triangle PQR in the y-axis.
Write down the coordinates of the image points.

2 Draw triangle ABC on squared paper. The coordinates are
A (−4,6)   B (5,3)   C (2,1)
Reflect triangle ABC on the x-axis.
Write down the coordinates of the image points.

CALCULATING ANGLES

Calculate the angles marked with a letter.

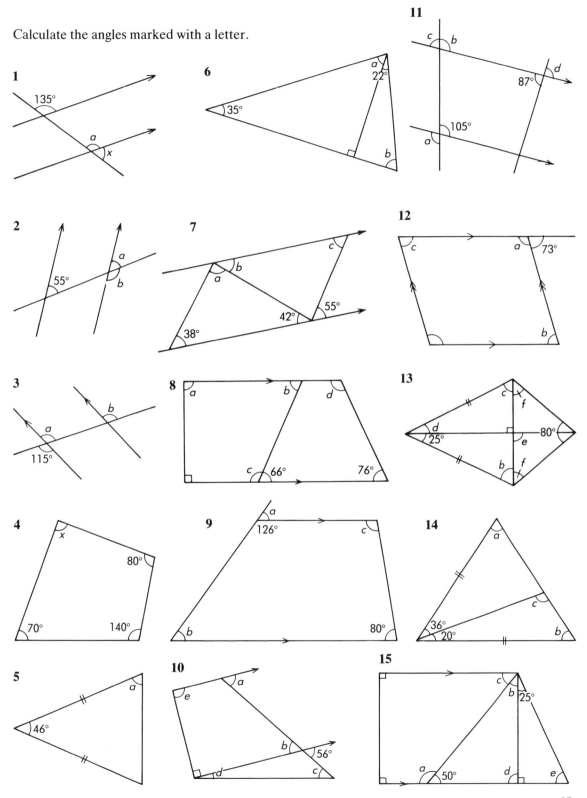

TANGENTS TO A CIRCLE

A 1 Describe each of the 3 diagrams so someone could draw each one without seeing this page.
2 Which one of these diagrams does not have line symmetry?
  Make a larger copy of the two diagrams with line symmetry.
  Show on them the facts associated with their symmetry.

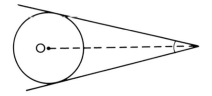

B In these diagrams O is the centre of each circle.
  For each question make a larger copy of the diagram and calculate the size of the marked
  ·angles. Write the answers in this way: $\angle APQ =$ °

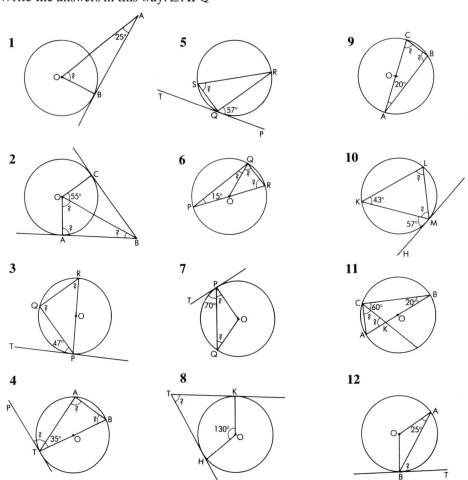

REFLECTIONS

EXAMPLE
The line AB has been reflected in M.
Its image is A′B′.

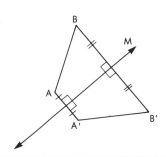

A (i) Copy the shape onto squared paper.
 (ii) Make a list of the coordinates of the points.
 (iii) Complete the shape so it has line symmetry about the *y*-axis.
 (iv) Make a list of the coordinates of the image points.
 (v) What pattern is there in the coordinates?
 (vi) Is the pattern in the coordinates the same for each shape?

B (i) Copy each shape.
 (ii) Reflect the shape in the balance line.
 (iii) Name any of the shapes which are familiar.
 (iv) Make a list of the facts which are shown by symmetry.

1

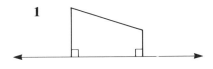

1

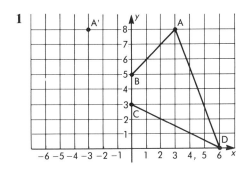

2

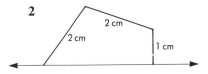

3

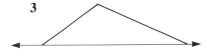

2

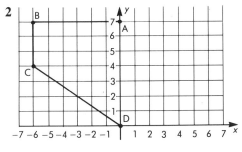

4

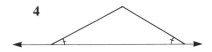

5

3

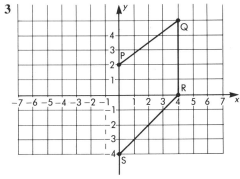

6

CONGRUENT SHAPES

EXAMPLE

These pairs of shapes are congruent.

P QRS
ADCB
shows that PQ fits AD, QR fits DC etc.

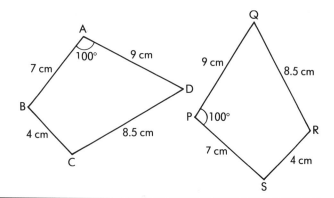

A Name the pairs of congruent shapes.

B Name the pairs of congruent shapes. Show how they fit together.

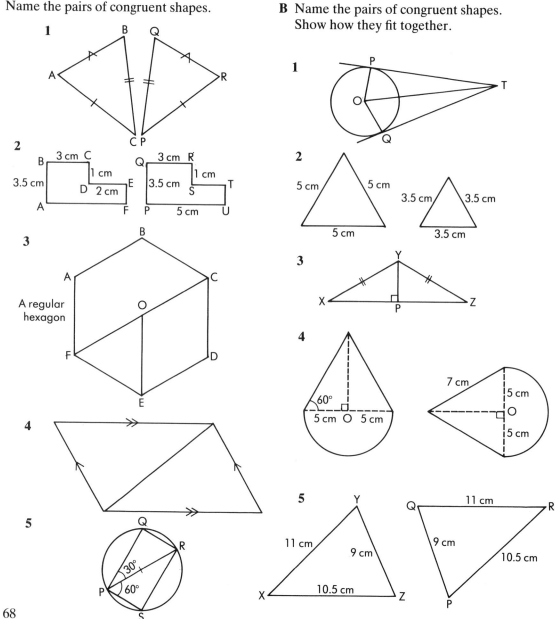

JIGSAWS AND SYMMETRY

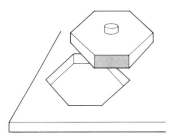

A This hexagonal piece is regular and fits neatly
into the space on the board.
One edge has been painted red to identify it.

1 In how many different positions can this red edge be placed into the space?
2 What information does this give about the symmetry of a regular hexagon? This diagram
may help the explanation:
3 Through how many degrees must the piece be turned to move the red edge from one position
to the next?
4 On the board there is another space which has not
been drawn. This hexagonal piece just fits that space.
In how many ways can it be placed into its space?
What facts could be learned from this?

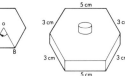

B

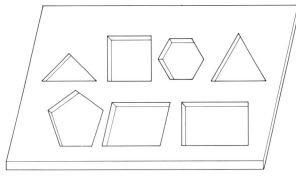

This is the base board of another jigsaw.
One edge of each piece has been marked.
Copy and complete this table:

Name of shape	Number of sides	Rotational symmetry	Size of each angle
Hexagon			
Square			

C These three-dimensional shapes have symmetry.
Group them according to their symmetry.
How is symmetry useful in everyday life?

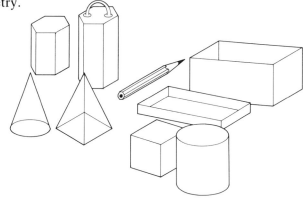

REGULAR POLYGONS

EXAMPLE

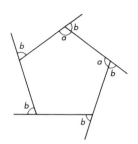

The exterior angles add up to 360°

$5b = 360$ so $b = 72$
$a + 72 = 180$
$a = 108$

A regular shape has equal sides and equal angles.

A Copy each regular polygon.
Go round each shape marking
the exterior angles.

1

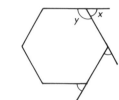

Calculate x and y in this hexagon.

2

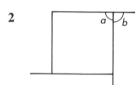

Calculate a and b in this square.

3

Calculate p and q in this
equilateral triangle.

4

Calculate x and y in this octagon.

5

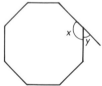

Calculate x and y in this pentagon.

B All these figures are regular.

1

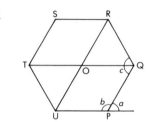

(i) Name the shape PQRSTU.
(ii) Calculate a, b, c.
(iii) Name triangle OQR.
(iv) Calculate angle UOQ.
(v) Name the figure PQOU.
(vi) Name a figure congruent to PQOU.

2

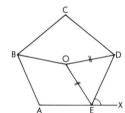

(i) Name this shape.
(ii) Calculate angle DEX.
(iii) Calculate angle AED.
(iv) OE bisects angle AED.
Calculate angle OED.
(v) Calculate angle ODE.
(vi) Name triangle ODE.
(vii) Name a shape congruent to ABOE.

3

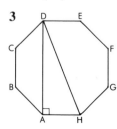

(i) Name this shape.
(ii) Name a shape congruent to
ABCDH.
(iii) Calculate angle AHG.
(iv) Calculate angle AHD.
(v) Calculate angle ADH.

TESSELLATING SHAPES

EXAMPLE

A tessellation is a tiling pattern which could go on forever.
The basis of this pattern is a square.

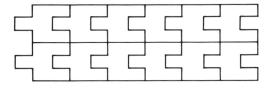

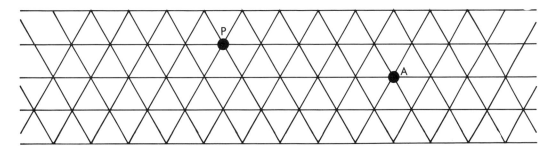

1 In this tessellation, how many triangles meet at P?

2 Copy and complete this diagram of the point P.

What is the total of the angles at P?
How big is each angle?

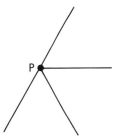

3 Name the shape formed when 6 equilateral triangles are grouped round a point.

4 How many of the new shapes meet round the point A?

5 What is the size of each interior angle of the new shape?

6 Name the shape formed when 2 equilateral triangles are put together.
Make a tessellation using this shape.

7 Name the shape formed by three equilateral triangles.
Will this shape tessellate?

8 Use different grid papers to invent some tessellating patterns.
Try changing the basic shape by removing a piece from one side and putting it on the opposite side.

11 trigonometry

Any triangle

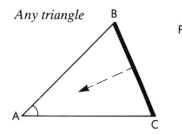

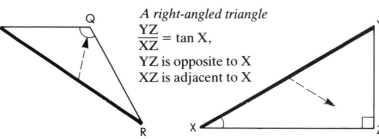

In △ ABC, the side opposite to ∠A is BC.
In △ PQR, the side PR is opposite to ∠Q.

A right-angled triangle

$$\frac{YZ}{XZ} = \tan X,$$

YZ is opposite to X
XZ is adjacent to X

In △ XYZ, the side opposite to ∠Z is XY.
(i) XY is the hypotenuse because it is opposite to the right angle.
(ii) XZ is adjacent to ∠X because it is next to ∠X.

A 1 Measure the sides and angles of these triangles.
Make tables like this to show the measurements of each triangle:

	Side	Opposite angle
longest	DF	
shortest		

2 The sides have been put in order of length.
Is there any order in the angles?
Write a rule about the shortest side of a triangle and the side opposite to it.

B These are approximate values for tan:
tan 20° = 0.36   tan 25° = 0.47
tan 30° = 0.58   tan 35° = 0.70
tan 40° = 0.84   tan 45° = 1.00
tan 50° = 1.19   tan 55° = 1.43
tan 60° = 1.73   tan 65° = 2.14

1 If tan A = 0.36, find ∠A
2 If tan B = 0.7, find ∠B
3 If tan A = 1.00, find ∠A
4 If tan E = 2.14, find ∠E
5 If tan X = 1.19, find ∠X
6 If tan H = 0.84, find ∠H
7 If tan Y = 1.73, find ∠Y
8 If tan X = 0.47, find ∠X
9 If tan X = 1.43, find ∠X
10 If tan P = 0.58, find ∠P

C 1

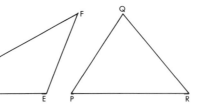

AB = 2.4 cm
BC = 5 cm
Find ∠C

2

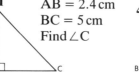

PQ = 1.2 cm
QR = 2 cm
Find ∠R

3

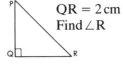

XZ = 8 cm
YZ = 8 cm
Find ∠X

4

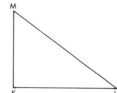

AC = 6.9 cm
BC = 4 cm
Find ∠B

5

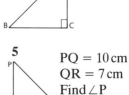

PQ = 10 cm
QR = 7 cm
Find ∠P

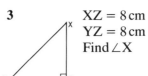

EXAMPLES

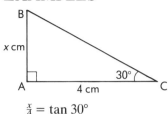

$\frac{x}{4} = \tan 30°$

$x = 4 \times \tan 30°$

$= 4 \times 0.58$

$AB \simeq 2.3 \text{ cm}$

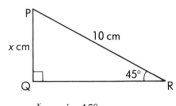

$\frac{x}{10} = \sin 45°$

$x = 10 \times \sin 45°$

$= 10 \times 0.71$

$PQ \simeq 7.1 \text{ cm}$

$c^2 = a^2 + b^2$

$c^2 = 7^2 + 8^2$

$= 49 + 64$

$= 113$

$c \simeq 10.6 \text{ cm}$

A Use these approximate values

tan 20 = 0.36	tan 48 = 1.11
tan 31 = 0.60	tan 53 = 1.33
tan 36 = 0.73	tan 58 = 1.60
tan 42 = 0.90	tan 60 = 1.73

B Use these approximate values

sin 25 = 0.42	sin 45 = 0.71
sin 30 = 0.5	sin 53 = 0.80
sin 39 = 0.63	sin 57 = 0.84
sin 44 = 0.69	sin 60 = 0.87

C Use a calculator:

1
$BC = 5 \text{ cm}$
$\angle C = 20°$
Find AB

2
$QR = 10 \text{ cm}$
$\angle R = 42°$
Find PQ

3
$XY = 2 \text{ cm}$
$\angle X = 31°$
Find YZ

4
$EF = 3 \text{ cm}$
$\angle E = 53°$
Find DF

5
$XY = 5 \text{ cm}$
$\angle X = 36°$
Find YZ

6
$PQ = 4.5 \text{ cm}$
$\angle R = 48°$
Find PQ

7
$BC = 6.2 \text{ cm}$
$\angle C = 60°$
Find AB

8
$DE = 10.1 \text{ cm}$
$\angle D = 58°$
Find EF

1
$XZ = 10 \text{ cm}$
$\angle Z = 25°$
Find XY

2
$AC = 4 \text{ cm}$
$\angle A = 44°$
Find BC

3
$PR = 3.5 \text{ cm}$
$\angle R = 30°$
Find PQ

4
$DF = 7.1 \text{ cm}$
$\angle F = 39°$
Find DE

5
$XY = 4.8 \text{ cm}$
$\angle Y = 45°$
Find XZ

6
$AC = 9 \text{ cm}$
$\angle A = 60°$
Find BC

7
$XZ = 8 \text{ cm}$
$\angle Z = 53°$
Find XY

8
$RQ = 12 \text{ cm}$
$\angle R = 57°$
Find PQ

1
$a = 3 \text{ cm}$
$c = 4 \text{ cm}$
Find b

2
$p = 5 \text{ cm}$
$r = 12 \text{ cm}$
Find q

3
$a = 8 \text{ cm}$
$b = 9 \text{ cm}$
Find c

4
$b = 7 \text{ cm}$
$c = 9 \text{ cm}$
Find a

5
$p = 2.5 \text{ cm}$
$r = 6 \text{ cm}$
Find q

6
$y = 6 \text{ cm}$
$z = 8 \text{ cm}$
Find x

7
$a = 10 \text{ cm}$
$b = 24 \text{ cm}$
Find c

8
$a = 5 \text{ cm}$
$b = 7 \text{ cm}$
Find c

1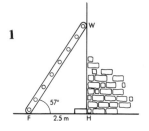

A ladder FW is leaning against a wall. The foot F of the ladder is 2.5 m from the wall. The ladder is sloping at an angle of 57°.
Calculate HW.
What does the length HW represent?
(sin 57° = 0.84, tan 57° = 1.54)

2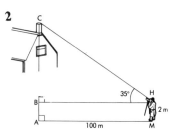

A man standing at M looks through an instrument at the top of a chimney. The instrument tells him the angle of elevation is 35°. He is standing 100 m from the base of the chimney.
Sketch △HBC and calculate the length of BC.
What is the height of the top of the chimney?
(sin 35° = 0.57, tan 35° = 0.7)

3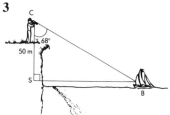

A man is standing at C on the top of a cliff which is known to be 50 m above sea level. He looks at the boat B through the instrument which measures the angle. It is 68°.
Sketch △SCB and calculate the length SB.
How far is the boat from the base of the cliff?
(sin 68° = 0.93, tan 68° = 2.48)

4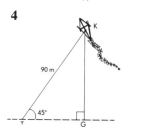

A boy is flying his kite on 90 m of string. The string is sloping at 45° to the ground.
Sketch △TGK and calculate the length of KG.
What does KG represent?
(sin 45° = 0.71, tan 45° = 1)

5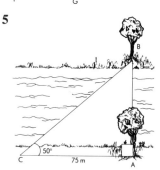

The two trees at A and B are on opposite sides of the river. The line AB is at right angles to the bank of the river.
A man standing at C is 75 m from A. He sights the tree at B and finds the angle is 50°.
Sketch △ABC and calculate the length of AB.
What does the length AB represent?
(sin 50° = 0.77, tan 50° = 1.19)

6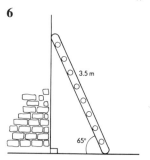

A ladder 3.5 m long is to be used against a wall. When it is sloping at 65° how far up the wall does it reach?
(sin 65° = 0.91, tan 65° = 2.14)

12 algebra

EXAMPLE

If $a = 3$, $b = 7$, $c = 0$, $d = -2$

$ab = 3 \times 7 = 21$	$5c = 5 \times 0 = 0$	$5d = 5 \times (-2) = -10$
$2ab = 2 \times 3 \times 7 = 28$	$b^2 = 7 \times 7 = 49$	$bd = 3 \times (-2) = -6$
$5b = 5 \times 7 = 35$	$4a^2 = 4 \times 3^2 = 36$	$d^2 = (-2) \times (-2) = 4$
	$5(2a + b) = 5(6 + 7)$	
	$= 5 \times 13 = 65$	

A If $p = 2$, $q = 5$, $r = 10$
calculate the value of:

1 $5p$	**11** pq	
2 $4q$	**12** pr	
3 $3r$	**13** qr	
4 $6p$	**14** rq	
5 q^2	**15** $4pq$	
6 r^2	**16** $3pr$	
7 p^2	**17** $5rp$	
8 $3q$	**18** $6rq$	
9 $7r$	**19** $3r^2$	
10 $3p$	**20** $2q^2$	

B If $a = 3$, $b = 4$, $c = 1$
find the value of:

1 $3b + c$	**11** $2(a + b)$
2 $2a + b$	**12** $5(b - c)$
3 $c + 2a$	**13** $3(2b + a)$
4 $5c + a$	**14** $4(b - a)$
5 $2c + 5b$	**15** $4(b - 2c)$
6 $7a - 5b$	**16** $10(a - 2)$
7 $4 + 3a$	**17** $7(1 + 3a)$
8 $5c + 2$	**18** $5(6 - b)$
9 $3b - 3$	**19** $4(9 - 2b)$
10 $2a - 2$	**20** $5(2a + 1)$

C If $x = 5$, $y = 3$, $z = -2$
find the value of:

1 x^2y	**11** $x + z$
2 xy^2	**12** $z + 7$
3 $(xy)^2$	**13** $3z$
4 $(3x)^2$	**14** $4z + 2x$
5 $y^2 + 1$	**15** $5 + z$
6 $x^2 - 4$	**16** z^2
7 $2x^2$	**17** x^3
8 $1 + 3y^2$	**18** y^3
9 $15 - 3x$	**19** z^3
10 $y + z$	**20** xyz

D 1 If $a = 0.7$, $b = 5.6$, $c = 4.9$, use a calculator to find the value of
(i) $9ab$, (ii) $a(b + c)$, (iii) $b^2 + 5c$ and (iv) $a + \frac{b}{c}$.

2 A manufacturer makes a product from 2 materials which cost £c per ton and £b per ton.
The cost of the materials varies, so he uses the formula £$(0.15c + 0.35b)$ to find how much 1 ton of the product has cost.
When $c = 49$ and $b = 37$, what was the cost of the materials for 1 ton of the product?

3 A firm has to deliver three types of parcels. The packing department uses red, green and blue labels to show the different types.

Weight	Label colour	Number of parcels
45 kg	red	r
37 kg	green	g
28 kg	blue	b

To calculate the weight of a load the formula $(45r + 37g + 28b)$ kg is used.
When there are 7 red parcels, 12 green parcels and 6 blue parcels,
what is the total weight of the load?

EXAMPLE

$x - 4.7 = 5$
$\quad x = 5 + 4.7$
$\quad x = 9.7$

$4x + 1 = 11$
$\quad 4x = 10$
$\quad x = 2\frac{1}{2}$

$3x^2 = 12$
$\quad x^2 = 4$
$\quad x = 2$

A Find the value of x:

1 $x + 3 = 9$
2 $5 + x = 12$
3 $x - 2 = 8$
4 $x - 3 = 4$
5 $5 - x = 2$
6 $8 + x = 11$
7 $x - 4 = 9$
8 $x + 7 = 12$
9 $x + 1 = 3.5$
10 $x + 2 = 7.3$
11 $x - 2 = 5.8$
12 $x - 3 = 4.9$
13 $x - 2.3 = 5$
14 $x - 3.1 = 6$
15 $x + 4 = 7.5$
16 $1.3 + x = 8$
17 $4 - x = 1$
18 $7 - x = 2.5$
19 $5 - x = 1.4$
20 $8 + x = 10.8$

B Solve:

1 $2x = 6$
2 $3x = 15$
3 $2x = 7$
4 $3x = 10$
5 $4x = 12$
6 $4x = 10$
7 $5x = 20$
8 $3x = 7$
9 $6x = 24$
10 $2x = 9$
11 $3x + 1 = 10$
12 $2x - 1 = 9$
13 $1 + 5x = 21$
14 $4x + 1 = 3$
15 $6x + 4 = 19$
16 $3 + 8x = 5$
17 $10x - 3 = 12$
18 $2x - 7 = 4$
19 $1 + 5x = 7$
20 $2x - 5 = 2$

C Find the value of x:

1 $x^2 = 9$
2 $x^2 = 16$
3 $x^2 = 36$
4 $x^2 = 49$
5 $x^2 = 100$
6 $x^2 = 25$
7 $x^2 = 4$
8 $x^2 = 64$
9 $x^2 = 81$
10 $x^2 = 0$
11 $x^2 = 1$
12 $x^2 + 1 = 10$
13 $x^2 - 1 = 3$
14 $x^2 + 5 = 30$
15 $x^2 + 6 = 15$
16 $x^2 - 4 = 60$
17 $x^2 + 9 = 90$
18 $2x^2 = 8$
19 $3x^2 = 48$
20 $2x^2 = 50$

D

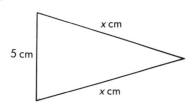

1 (i) Write down a formula for the perimeter of this triangle.
   (ii) If the perimeter of this triangle is 14 cm, write an equation involving x.
   (iii) Use this equation to find the value of x.

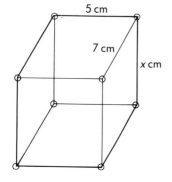

2 The diagram shows a framework made from thin rods. The measurements of the top are fixed as 7 cm by 5 cm.
   (i) Make a sketch of the top of this framework showing its measurements.
   (ii) What is the perimeter of the top?
   (iii) What is the total length of the rods used for the base?
   (iv) What is the total length needed for all the upright pieces?
   (v) If the total length of all the edges must be 60 cm write an equation and solve it to find x.
   (vi) What is the length of each vertical rod?

EXAMPLE

$2a + 3a + 1 = 5a + 1$

$5x + 1 + x = 6x + 1$

$4x - x = 3x$

$3k + h - k = 2k + h$

A Simplify:

1 $4a + 3a$
2 $5s + s$
3 $6x - 2x$
4 $5n - 4n$
5 $3x - 2x$
6 $4d + 3d$
7 $5y - 4y$
8 $2k + 3k$
9 $5x + 4x$
10 $4p + p$
11 $3r + 1 + 2r$
12 $x + 3 + 2x$
13 $4y + 5 + y$
14 $5d - d + 3$
15 $6 + 4f - 3f$
16 $2h + 4 - 2h$
17 $4x - 3x + 6$
18 $4y - y + 1$
19 $2k - 5 + k$
20 $3d + d - 3$

B Simplify:

1 $4x + y - 2x$
2 $3a + 2b + a$
3 $4d - 3 + d$
4 $5k - h + 2k$
5 $6 - 2x + 4$
6 $2x + 3 + 3x$
7 $4y + x + 2y$
8 $3a + 2b + a$
9 $5x - 2x - y$
10 $6a + b - 2a + b$
11 $4x + 2y - 4x$
12 $3k - 2k + 5h$
13 $7 - 2x + 5 + 3x$
14 $3y - 5 + y + 6$
15 $5a - 7 + 2a + 8$
16 $7 - 2d + 3 + 3d$
17 $6x + 4 - x - 4$
18 $5k - 3h + 7h - k$
19 $x + y + 2x + 3y$
20 $2a - b - 3a + b$

C Simplify:

1 $a + 3a + 4$
2 $2x + x + 6$
3 $4b - b + c$
4 $5y - k - y$
5 $4k - k + 3h$
6 $7 + 3a$
7 $5x + 3y - y$
8 $4p + p + q$
9 $5x + 2y - y$
10 $c + 4c + 3$
11 $a + b + 3a$
12 $p + k + 2p$
13 $2x + y + 3x$
14 $7k + h + 2k + 3h$
15 $4x - 3y + 2x + 5y$
16 $4d - c + 2c - d$
17 $3b + 4a - 2b + a$
18 $6k - 4a + 3k + 7a$
19 $6x + 2 - 4a + x$
20 $7 - 5x + 8x + 2$

D 1

The sides of this rectangle are $2x$ cm and x cm.
Write down the formula for its perimeter.

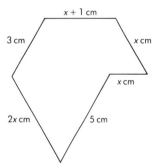

2 The sides of this shape are $(x + 1)$ cm, x cm, x cm, $2x$ cm,
5 cm and 3 cm long. Write down a formula for its perimeter.

3

One side of this rectangle is 5 metres long and
the other side is $(x + 1)$ metres.
Write down a formula for its area.

4 Ann has N marbles and Joan has twice as many as Ann. Write a formula for the number Joan has.
Kathy has 3 more than Ann. Write a formula for the number Kathy has.
Write a formula for the number of marbles they have altogether.

MACHINES AND FORMULAE

A Two numbers are put into
this 'add machine':

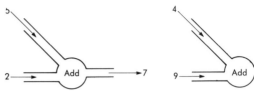

   1 What comes out this time?
   2 What comes out if you put 6 and 3 in?
   3 If x and y are entered and z comes out, write this as a formula.

B **1** What does this machine do to numbers?
   2 Put 7 into this machine. What comes out?
   3 If a is put into the machine and x comes out, write this as a formula.

C Machines can be put together like this:
   1 Put 4 and 5 in. What comes out?
   2 Use brackets to show how 4 and 5 became 18
as they went through the machine.
   3 If y is the outcome when a and b are entered in this machine write
this as a formula.

D **1** This is a 'perimeter machine':
What answer does it give?

Use the 'perimeter machine' to find the distance all round these rectangles:

2   **3**   **4**   **5**

   6 Would this machine give the correct result for the perimeter of this shape?

E Write a formula for each of these machines:

   1

   2

   3

   4

F

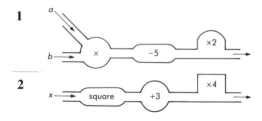

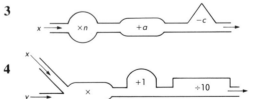

1 This set of machines works out the
weekly paper bill. Put in 75p for the
cost of daily papers and 240p for the
cost of Sunday papers and magazines.
What is the unit of the outcome?

78

2 Here is another set of machines. They work out hire purchase.

£60 → ÷10 → ×11 → ÷12 → amount to pay each month

Put in £60. Does £5.50 come out?
Put in £960. How much does the machine say has to be paid each month?
Put in £840. How much does the machine say has to be paid each month?
Does this machine take account of a deposit or down payment?

A 1 When $x = 1$, what is y?
  2 What are the coordinates of the point when $x = 1$?
  3 Give the coordinates of the points:
    (i) when $x = 3$
    (ii) when $x = -3$
  4 When $y = 6$, what is x?
  5 When $x = -1$, what is y?
  6 When $y = 10$, what is x?
  7 When $y = -12$, what is x?

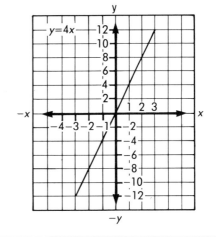

B 1 Write down the coordinates of these points:
    (i) when $x = 4$
    (ii) when $x = 6$
    (iii) when $x = 10$
  2 When $x = 3$, what is y?
  3 When $x = 7$, what is y?
  4 When $y = 2\frac{1}{2}$, what is x?
  5 When $y = 5\frac{1}{2}$, what is x?
  6 When $y = 2$, what is x?

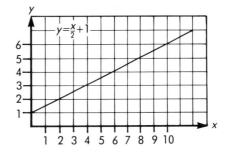

C 1 Write down the coordinates of these points:
    (i) when $x = 2$
    (ii) when $x = -1$
  2 What value of x makes $y = 2$?
  3 What is the value of x when $y = 2\frac{1}{2}$?
  4 What is x, when $y = -1$?
  5 As x increases, what happens to y?
  6 Use the graph to solve the equation
    $-3 = 3 - x$

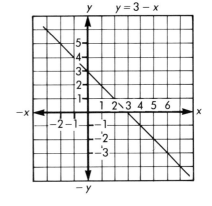

EXAMPLE

$x + 5x = 6x$

$4a - a = 3a$

$5(4x - 2) = 20x - 10$

$-3k + 5k = 2k$

$$\underline{\quad}|\underline{\quad}|\underline{\quad}|\underline{\quad}|\underline{\quad}|\underline{\quad}|\underline{\quad}$$
$$n{-}2 \quad n{-}1 \quad n \quad n{+}1 \quad n{+}2 \quad n{+}3$$

A Simplify:

1 $2(3a + 5)$
2 $4(3x - 1)$
3 $3(a - b)$
4 $4(2x + y)$
5 $5(k - 2)$
6 $3(4a + b)$
7 $5(3a - 1)$
8 $2(3y - 4)$
9 $3(5 - 2a)$
10 $4(1 - x)$
11 $8(a + 2b)$
12 $3(2 - x)$
13 $5(x + 4)$
14 $6(2 + x)$
15 $3(a - 2)$
16 $4(x + y)$
17 $2(3 - x)$
18 $4(b - 2a)$
19 $2(7 + 3a)$
20 $5(2x + y)$

B Simplify:

1 $3(2x + y) + y$
2 $5(3x - 1) + 8$
3 $2(x + y) + 2x$
4 $4(2a - 3) + a$
5 $3(3h + 1) - 9h$
6 $2(4 + 2d) + d$
7 $3(2 - 3x) + 12x$
8 $3(4d - c) + d$
9 $2(x - 1) - x$
10 $2(x - 3) + 6$
11 $2(x + y) + 3(x + 2y)$
12 $3(a + b) + 2(a - b)$
13 $2(3x - 2) + 5$
14 $7 + 3(3x - 2)$
15 $5 + 3(2 - x)$
16 $8 + 2(x - 4)$
17 $4x + 2(y - x)$
18 $3y + 3(2x - y)$
19 $2x + 5(a - x)$
20 $x + 2(3 - 4x)$

C Write the number:

1 4 more than x
2 2 less than x
3 7 more than x
4 5 more than a
5 3 less than y
6 1 more than k
7 2 more than y
8 2 less than k
9 3 more than x
10 5 more than k
11 3 times a
12 4 times x
13 10 times k squared
14 4 times a squared
15 7.3 times y
16 0.4 times p
17 $\frac{1}{4}$ of y
18 $\frac{1}{2}$ of h
19 $\frac{1}{10}$ of y
20 $\frac{1}{5}$ of x

D **1** The formula for this sequence is $x + 3.5$: 5, 8.5, 12, 15.5, ...
Write the next 3 numbers in this sequence.

2 The formula for this sequence is $x - \frac{1}{2}$: 10, $9\frac{1}{2}$, 9, $8\frac{1}{2}$, . . .
Write the next 3 numbers in the sequence.

3 The formula for this sequence is $3t$: 4, 12, 36, . . .
Write the next number in the sequence.

4 The formula for this sequence is $\frac{1}{2}k^2$, where $k = 1$, 2, 3, . . . : $\frac{1}{2}$, 2, $4\frac{1}{2}$, 8, $12\frac{1}{2}$, ...
Write the next two numbers in the sequence.

5 The formula for this sequence is $k^2 + 1$, where $k = 2$, 3, . . . : 5, 10, 17, 26, ...
What is the next number in this sequence?

FORMULAE

EXAMPLES

$5x$ means 5 times x
so x is $\frac{1}{5}$ of $5x$
  or x is $5x \div 5$

If $5x = a$
then $x = \frac{1}{5}$ of $a = \frac{a}{5}$
check: if $a = 10$,
then $\frac{1}{5}$ of $10 = 2$

$x + 3$ means 3 more than x
so if $x + 3 = k$
then $x = k - 3$
If $\frac{x}{7} = d$
then $x = 7d$

A Make x the subject:

1. $3x = a$
2. $2x = k$
3. $x + 1 = b$
4. $x + 5 = p$
5. $2x = n$
6. $x - 2 = d$
7. $x - 3 = a$
8. $x + 3 = a$
9. $2 + x = v$
10. $5 + x = t$
11. $4x = k$
12. $3x = p$
13. $x - 5 = t$
14. $10x = 3d$
15. $4x = k$
16. $x + 4 = 3t$
17. $x - 3 = d$
18. $5x = a$
19. $10x = t$
20. $5 - x = k$

B Make x the subject:

1. $n = \frac{x}{3}$
2. $k = \frac{x}{5}$
3. $a = \frac{x}{4}$
4. $\frac{x}{2} = h$
5. $\frac{x}{10} = t$
6. $\frac{x}{4} = k$
7. $\frac{x}{3} = h$
8. $n = \frac{x}{7}$
9. $d = \frac{x}{5}$
10. $\frac{x}{4} = t$
11. $3x = p$
12. $5x = k$
13. $n = 10x$
14. $\frac{x}{2} = N$
15. $A = 3x$
16. $P = 4x$
17. $E = 12x$
18. $H = 5x$
19. $p = 6x$
20. $J = 5x$

C Make x the subject:

1. $x + 1 = k$
2. $2x + 1 = a$
3. $x - 2 = n$
4. $3x - 2 = t$
5. $1 + 5x = n$
6. $5x - 4 = b$
7. $4x - 2 = a$
8. $10x + 3 = t$
9. $6x - 1 = h$
10. $3a + 4x = 5a$
11. $x + a = k$
12. $x + 2p = q$
13. $x + h = 3b$
14. $x + 2h = g$
15. $x + 5t = s$
16. $x - a = 2b$
17. $x - n = d$
18. $x + 5c = t$
19. $x + k = a$
20. $h + x = d$

D These shapes are regular. Write down a formula for the perimeter of each:

1

2

3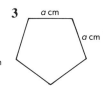

4

5

Write down a formula for the perimeter of these shapes:

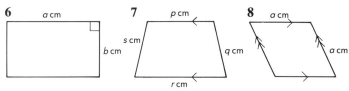

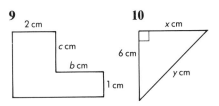

81

FACTORS AND PRODUCTS

EXAMPLE

ax means a times x
and $a(x + 2)$ means a times $(x + 2)$

$kx + ka$ means $k \times x \ + \ k \times a$
The factor k is common to both terms
$kx + ka = k(x + a)$
$x^2 + 5x = x(x + 5)$

A Simplify:

1 $4(x - 1)$
2 $a(x + 3)$
3 $a(x - 5)$
4 $a(1 - x)$
5 $b(x - 2)$
6 $k(x + 1)$
7 $x(2 + n)$
8 $x(5 - a)$
9 $x(t + 4)$
10 $a(x + c)$
11 $t(2 - x)$
12 $4(x + b)$
13 $3(x + c)$
14 $7(d - x)$
15 $x(3 + x)$
16 $x(5 + x)$
17 $a(7 + x)$
18 $d(c + x)$
19 $k(x + a)$
20 $p(x - d)$

B Find the common factor, then write as a product.

1 $5x + 5a$
2 $3x - 3b$
3 $ax + 2a$
4 $bx + 3b$
5 $kx - 5k$
6 $7d - 7x$
7 $ax + 3a$
8 $bx - 3b$
9 $kx + 4k$
10 $4n + nx$
11 $3a + ax$
12 $5c + cx$
13 $7k - kx$
14 $ax + 2x$
15 $5x + ex$
16 $4x + ax$
17 $10x - 10a$
18 $8x - 12a$
19 $6x - 3b$
20 $10a + 15x$

C Write as products.

1 $x^2 + 2x$
2 $x^2 + bx$
3 $3x + x^2$
4 $5y + y^2$
5 $y^2 - 6y$
6 $x^2 - 3x$
7 $x^2 + ax$
8 $y^2 + 7y$
9 $x^2 - bx$
10 $x^2 - 8x$
11 $a^2 + 2a$
12 $t^2 - 5t$
13 $d^2 + 3d$
14 $5a + a^2$
15 $c^2 - 5c$
16 $k^2 + 10k$
17 $4t + t^2$
18 $d^2 - 8d$
19 $4a + a^2$
20 $7k + k^2$

D Make an approximation using whole numbers to check some of the answers.
Show the numbers used for these checks.
Use a calculator to find the exact value of:

1 $2(3.4 + 1)$
2 $5(6.3 - 4)$
3 $7(8.1 + 2)$
4 $5(6.3 - 2.9)$
5 $4(7.8 + 1.4)$
6 $15(6.3 + 4)$
7 $23(5 + 6.7)$
8 $31(8 + 0.7)$
9 $53(9 + 0.4)$
10 $26(7 + 1.8)$

11 $4(3.6 + 1) + 3$
12 $5(7.2 - 1) - 1$
13 $2(4.8 + 3.7) + 5$
14 $6(3.9 + 0.3) + 4$
15 $10(4.4 - 3.9)$
16 $4 + 8(3.6 + 2.1)$
17 $7 + 9(5.1 - 2.9)$
18 $8 - 3(2.7 - 1.5)$
19 $12 - 6(7.3 - 6.8)$
20 $10 + 3(5.1 - 3.8)$

21 4.7^2
22 5.9^2
23 3.95^2
24 $3(4.1)^2$
25 $2(5.3)^2$
26 $7(5.9)^2$
27 $8(4.7)^2$
28 $9(0.4)^2$
29 $7(0.5)^2$
30 $8(0.61)^2$

EXAMPLE

$2(x-3) = 12$	$\frac{x}{5} = 10$	$\frac{15}{y} = 5$
$x - 3 = 6$	$x = 50$	$y = 3$
$x = 9$		

A Find the value of x:

1 $2(x - 4) = 10$
2 $3(x + 2) = 21$
3 $3(x + 1) = 12$
4 $2(1 + x) = 10$
5 $5(x + 1) = 30$
6 $4(2 + x) = 10$
7 $2(x - 1) = 9$
8 $3(3 + x) = 15$
9 $2(x - 5) = 7$
10 $4(x - 3) = 10$
11 $5(x - 3) = 15$
12 $6(x + 2) = 15$
13 $2(x - 5) = 17$
14 $4(x - 2) = 6$
15 $10(x + 2) = 25$
16 $5(x - 7) = 12$
17 $4(x - 4) = 10$
18 $6(x - 2) = 15$
19 $4(x - 5) = 14$
20 $10(1 + x) = 25$

B Solve:

1 $\frac{x}{4} = 6$
2 $\frac{x}{2} = 6$
3 $\frac{x}{3} = 5$
4 $\frac{y}{2} = 10$
5 $\frac{y}{3} = 9$
6 $\frac{y}{3} = 1.5$
7 $\frac{x}{2} = 6.3$
8 $\frac{x}{2} = 4.8$
9 $\frac{y}{5} = 2.1$
10 $\frac{x}{3} = 3.3$
11 $\frac{x}{4} = 2.8$
12 $\frac{24}{y} = 3$
13 $\frac{30}{x} = 10$
14 $\frac{20}{y} = 4$
15 $\frac{12}{x} = 4$
16 $\frac{16}{x} = 2$
17 $\frac{21}{y} = 7$
18 $\frac{15}{y} = 3$
19 $\frac{24}{x} = 6$
20 $\frac{36}{y} = 9$

C Solve:

1 $x + 3 = 9$
2 $x + 5 = 7$
3 $4 + x = 11$
4 $x - 5 = 4$
5 $x - 2 = 8$
6 $2x = 10$
7 $4x = 12$
8 $2x = 3$
9 $2x = 9$
10 $2x + 3 = 11$
11 $3x - 1 = 14$
12 $1 + 3x = 22$
13 $4 + x = 10.5$
14 $x + 3 = 7.2$
15 $2x + 3 = 6$
16 $x^2 = 9$
17 $5x - 2 = 5.5$
18 $3x - 4 = 8.6$
19 $\frac{x}{5} = 2$
20 $\frac{y}{2} = 4$

D Find the value of x by writing an equation and solving it for each of these:

1

2

3
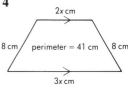

4
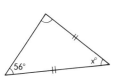

5
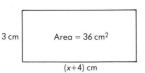

6

7
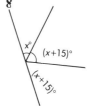

8
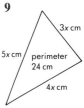

9

DIAGONALS OF A POLYGON

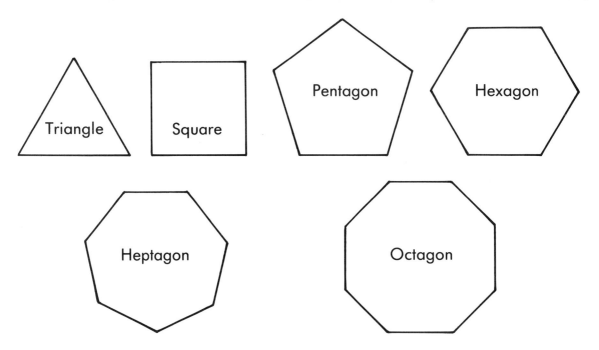

1 Make a large drawing of a pentagon.
  Draw all the diagonals and count them.
2 Draw a large hexagon and its diagonals.
3 Copy the table below.
4 Write in the table the number of diagonals in a pentagon and hexagon.
5 Which of these formulae gives the correct number of diagonals for a
  polygon with n sides: (i) $n + 3$,   (ii) $n - 3$ or (iii) $\frac{n}{2}(n - 3)$?
6 Use the formula to calculate the number of sides in more polygons.
7 There are more diagonals in a shape as the number of sides increases.
  How does the number of diagonals increase?
8 Are the same results true if the polygons are not regular?

Name of figure	Number of edges, n	Number of diagonals
Triangle	3	
Square		
Pentagon		
Hexagon		
Heptagon		
Octagon		
Nonagon		
Decagon		
15-sided polygon		
20-sided polygon		

13 probability and statistics

LUCK OF THE DRAW

If a bag contains five coloured balls (red, blue, black, white and green) the probability of drawing out the red ball is $\frac{1}{5}$. If the bag contains 2 red balls and 3 blue balls, there are two chances of drawing the red ball. So the probability of drawing the red ball is $\frac{2}{5}$.

A Find the probability of drawing from the bag:

1 2p from three 5p, two 2p coins
2 2p from three 5p, four 2p coins
3 5p from four 5p, six 2p coins
4 5p from three 5p, eight 2p coins
5 2p from nine 5p, four 2p coins
6 2p from five 5p, three 2p coins
7 5p from four 5p, seven 2p coins
8 2p from six 5p, eight 2p coins
9 2p from three 5p, nine 2p coins
10 5p from eight 5p, twelve 2p coins
11 a red ball from 3 red, 7 blue balls
12 a red ball from 5 blue, 2 red balls
13 a red ball from 4 blue, 1 red ball
14 a red ball from 3 black, 1 white, 3 red balls
15 a red ball from 6 green, 2 white, 3 red balls
16 a red ball from 4 blue, 3 white, 3 red balls
17 a red ball from 7 red, 7 white balls
18 a red ball from 5 red, 2 white, 3 blue balls
19 a red ball from 3 blue, 2 white, 2 red balls
20 a red ball from 4 green, 1 white, 1 red ball

B From a pack of 52 playing cards, find the probability of drawing:

1 a queen
2 the queen of diamonds
3 a red queen
4 a black king
5 an ace
6 the ace of hearts
7 a red card
8 a spade
9 the 10 of clubs
10 the 4 of hearts
11 the jack of spades
12 a three
13 a red four
14 the king of hearts
15 a diamond
16 the 7 of clubs
17 the 5 of diamonds
18 a red 2
19 a black king
20 the 6 of spades

C When a die is thrown, find the probability of getting:

1 5
2 an even number
3 1
4 a number greater than 4
5 2
6 an odd number
7 a number less than 4
8 3
9 an even number greater than 2
10 6

D Copy and complete this diagram to show what may happen when 3 coins are tossed together:

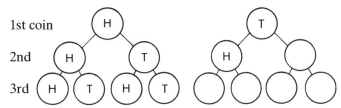

When 3 coins are tossed together, what is the probability of getting:

1 2 heads, 1 tail
2 1 head, 2 tails
3 3 heads
4 no heads
5 2 tails, 1 head
6 3 tails

1 Three letter nonsense words are to be made using the letters Q, K and S, e.g. QQQ.
   (a) How many words can be made so that all the letters in the word are the same?
   (b) If the first letter has to be Q, the next two letters may be any of the three letters, how many different words can be made?
   (c) How many different words can be made starting with K?
   (d) How many different words can be made from the letters Q, K and S?

2 Three digit numbers are to be formed using the digits 7, 5 and 2, e.g. 772.
   (a) How many of these numbers will be even?
   (b) How many different numbers can be formed?
   (c) If one of these numbers is chosen at random, what is the chance that it starts with 7?

3 Mary and Tom are playing a game with a 5-sided spinner numbered 0, 1, 2, 3 and 4.
   (a) When Mary spins it stops on 3. What is the probability that Tom will get a 3 when he spins it?
   They decide to play so that Mary spins twice and adds her two scores together. Then Tom does the same.
   (b) What is the highest score that Mary could get?
   (c) What is the lowest score that Tom could get?
   Complete this table to show all the possible totals

1st score		0	1	2	3	4
	0	0	1	2	3	4
	1	1	2			5
2nd	2					
	3					
	4		5		7	

   (d) What is the probability of Mary scoring a total of 0?
   (e) What is the probability of Tom scoring a total of 8?
   (f) What is the probability of Tom getting a total of 4?
   (g) What is the probability of Mary's total being an odd number?

4 A number is selected at random from the set 2, 3, 4, 5, 6, 7, 8, 9.
   (a) What is the probability that it is 5?
   (b) What is the probability that it is a prime?
   (c) What is the probability that it is a multiple of 3?
   (d) What is the probability that it is an even number?

5 A man has a tray of 50 young plants. He knows that the probability of each plant growing into a healthy plant is $\frac{7}{10}$. He can sell the healthy plants for 60p each. How much income can he expect?

EXAMPLE

The mean of 73, 48 and 44 is 45

$$\frac{73 + 48 + 44}{3}$$

The mean of 3 kg, 2.5 kg, 4 kg and 2.9 kg is 3.1 kg

$$\frac{3 + 2.5 + 4 + 2.9}{4}$$

Find the mean value of:

A
1. 3, 5, 10
2. 1, 1, 3, 4, 6
3. 10, 10, 12, 14
4. 65, 45, 30, 60
5. 7, 6.5, 10, 11
6. £4, £4.20, £4.10
7. 20p, 25p, 15p, 40p
8. 4p, 5p, 6p, 6p, 14p
9. £26, £31, £18
10. £4.50, £4.90

B
1. 3 mm, 4 mm, 4 mm, 5 mm
2. 1.5 km, 2.3 km, 2.5 km
3. 4 cm, 4 cm, 7 cm
4. 15 m, 15 m, 15 m, 21 m
5. 20 mm, 22 mm, 19 mm, 25 mm, 24 mm
6. 3 h, 3 h 20 min, 3 h 15 min, 3 h 3 min
7. 25 sec, 28 sec, 29 sec, 25 sec, 23 sec
8. 1 min, 55 sec, 58 sec, 1 min 7 sec
9. 12 sec, 12.5 sec, 12.4 sec
10. 18 sec, 19 sec, 19.5 sec, 17.5 sec, 18 sec.

C 1 Three boys' weights were 49 kg, 52 kg and 64 kg. What was their mean weight?

2 A shopkeeper's record of his daily takings is as follows:

Monday, £250;   Tuesday, £628;   Wednesday, £194;   Thursday, £938;
Friday, £1276;   Saturday, £1484.

(i) What was the total for the week?   (ii) What was the mean daily total?
If he had hoped for a mean daily taking of £800, how much more would he have needed to take in the week?

3 The lengths in millimetres of a sample of 10 bird's eggs were:
   20, 24, 22, 20, 21, 23, 22, 21, 22, 23.
   (i) Calculate the mean length of the eggs.   (ii) How many eggs were smaller than the mean?

4 As part of an experiment the heights of 5 tomato plants were measured.
   The fruit borne by each plant was weighed and recorded:

Height in m	0.9	0.65	1.35	0.65	0.9
Weight in kg	1.9	1.05	3.0	1.65	2.5

(i) Calculate the mean of the height of the plants.   (ii) Calculate the mean weight of fruit.

5 The table shows the number of days lost from work through illness in two workshops.

	Jan	Feb	Mar	Apr	May	June
Workshop A	32	28	12	9	15	6
Workshop B	45	36	10	3	2	0

Calculate the mean number of days lost per month for each workshop.
There are the same number of people in each workshop. Discuss the results.

ILLUSTRATING RECORDS

A The canteen supervisor kept a record of the number of broken plates:

	Monday	Tuesday	Wednesday	Thursday	Friday	Saturday	Sunday
Week 1		/		/	/	//	
Week 2		//		/		////	//

1 Copy and complete this histogram showing the breakages for the two-week period:

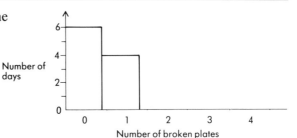

2 Find the mean number of broken plates.

3 On how many days was the number broken greater than the mean?

B A class of children were asked how many people live in their house.

1 How many homes had 3 occupants?
If both parents were in these homes, how many children are in a family?

2 How many homes had 5 occupants?
If both parents were in all these homes, how many children are there in each family?

3 Which is the largest family size?
How many families are this size?

4 Which family size occurs most often?

5 How many families are shown altogether?

6 What is the average number of occupants?

7 Copy and complete this table to show how many people lived in each house.

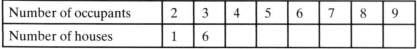

Number of occupants	2	3	4	5	6	7	8	9
Number of houses	1	6						

C A graph was made of the attendances at school made by a class.

1 How many half-day attendances are possible in a week?

2 How many children were not absent?

3 How many missed just one half-day?

4 How many were absent all the week?

5 How many children are in the class?

6 What was the total number of attendances?

7 What was the average number of attendances for each child?

8 What was the average attendance per session?

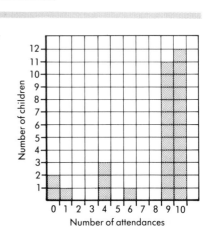

A Two dice are thrown together. The scores are added.

These dice show a score of 8.

What other combinations would give a score of 8?
Copy and complete this table to show all the possible scores:

	1	2	3	4	5	6
1						
2						
3						
4			8			
5						
6						

Write down the probability that the total shown on the two dice is:

1 6   **3** 12   **5** 10   **7** 7   **9** 4
2 8   **4** 5   **6** 3   **8** 0   **10** 1

B I am going on holiday on 4th July for a week to an area where the probability that it will rain on any particular day is $\frac{1}{10}$.

1 What is the probability that it will rain on 5th July?
2 What is the probability that it will rain on my last day?
3 What is the probability that it will be fine on 8th July?
4 What is the probability that two successive days will be fine?

C On a particular shooting range, the probability that I hit the target is $\frac{3}{4}$. My friend has a probability of $\frac{3}{5}$ that he hits the target. We make one shot each at the target.
1 What is the probability that I do not hit the target?
2 Find the probability that we both hit the target with that shot.
3 Find the probability that only one of us hits the target.

D Fred rates his chances of scoring a goal in a match as $\frac{1}{4}$.
Find the probability that in his next two matches:
1 he scores a goal in each.
2 he scores in only one of the matches.
3 he does not score in either.

E Jane is one of the children in a class of 20.
Five of the children are left handed and 7 children wear glasses.
What is the probability that
1 Jane is right handed?
2 Jane does not wear glasses?
3 Jane is left handed and wears glasses?
4 Which of the events above is most likely?

WINNING A COMPETITION

Plan a kitchen–HUGE PRIZES ! !

Choose the order you think most important for a new kitchen. Just write the letters.

A Eye level grill   F Refrigerator
B Extractor fan   G Dish washer
C Formica work surface H Tiled walls
D Washing machine
E Refuse disposal unit

First   Name..................
2nd   ....
3rd   Address................
4th   ....
5th   ....
6th   ....
7th   ....
8th

It is very unlikely that you will win this sort of competition. To find the reason you will need small pieces of paper labelled A, B, C, D, E, F, G, H.

Use the pieces marked A and B.
You can put them in order like this:

1 Use the pieces marked A, B and C.

C can be put into the [A/B] order in 3 places.

Use this to help you find all the different orders. Copy and complete it.

2 Use the pieces marked A, B, C and D.

In the first block A, B and C stay in order. There are four different spaces for D to go in to the

[A/B/C] order.

Copy and complete this. How many ways are there of putting four things in order?

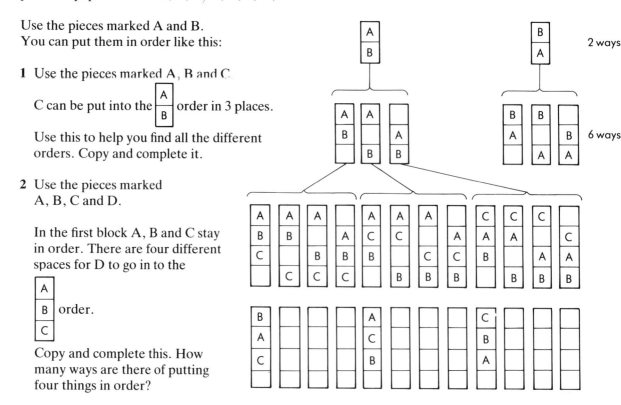

3 How many answers would not win first prize?

Copy and complete this table with your results:

Number of things to be put in order	1	2	3	4	5	6	7	8
Number of different orders	1	2	6					
Number of orders which won't win	0	1	5					

14 assorted situations

IN THE STOCKROOM

A

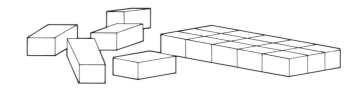

There has been a delivery of 70 identical packets.
The storekeeper starts to stack them away, making a bottom layer 5 packets long and 3 wide.

1 How many packets are in the bottom layer?

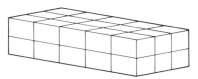

He makes another layer on top of the first.
2 How many packets have now been stacked?

3 When he has made 4 layers, how many packets have still to be stacked?
4 How many layers will he need altogether?

B The storekeeper has to stack a lot of cans into large cartons.
He has two sizes of carton. The inside measurements of one are
35 cm by 24 cm and 48 cm high.
The other measures 35 cm by 25 cm and is 24 cm high.
Which carton should he choose? Why?

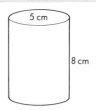

C These boxes are 3 cm wide, 5 cm high and 8 cm long.
Hundreds of them have to be packed into cartons.

These are the carton sizes available:

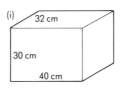

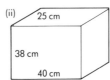

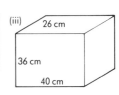

What is the largest number each carton would hold?
Which is the best size to choose?
Why?

D Eggs are packed on trays which hold 5 rows of 6.
1 How many dozen is this?
The full trays are stacked 6 deep in a large box.
The box is made to take 2 stacks of trays.
2 How many eggs are packed into one box?
3 If each egg weighs about 2 oz, make a statement about the weight of the box when it is full. (16 oz = 1 lb)

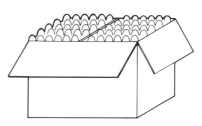

PAINTING CUBES

A

The 2-cube

How many small cubes are used to build this 2-cube?
Imagine that the outside of this 2-cube is painted.
How many small squares will have to be painted?
When it is knocked down some faces of the small cubes
have not been painted.
How many of the cubes have three faces painted?
Are any of the cubes unpainted?

B

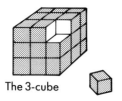

The 3-cube

Why is this called a 3-cube?
How many small cubes are used to build the bottom layer?
How many cubes are used to build the middle layer?
How many cubes are used altogether?

C

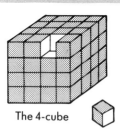

The 4-cube

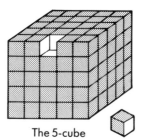

The 5-cube

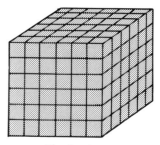

The 6-cube

The outside of the 6-cube is painted.
Then it is knocked down.

How many cubes have 2 faces painted?
How many cubes have 3 faces painted?
How many cubes have 1 face painted?
Are any of the cubes unpainted?

Copy and complete this table with your results:

	Number of small cubes	Number with 2 faces painted	Number with 3 faces painted	Number with 1 face painted	Number with 0 faces painted
2-cube	8	0	8	0	0
3-cube					
4-cube					
5-cube					
6-cube					

Can you see any patterns in the results?   Use the patterns to work out the 7-cube.

PYRAMIDS AND PRISMS

A The pyramid on a square base has 5 faces. Four of the faces
are triangles and the fifth is a square.
Check that this pyramid has 5 corners and 8 edges.

Look carefully at these pyramids and count their faces, corners, edges. Enter the results in a
table like the one at the foot of this page.

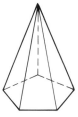
pyramid on a
pentagonal base

pyramid on a
hexagonal base

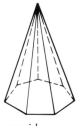
pyramid on a
heptagonal base

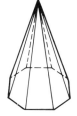
pyramid on an
octagonal base

B This prism has 2 triangular faces and 3 of the faces
are rectangles. It has 5 faces altogether.
Check that this prism has 9 edges and 6 corners.

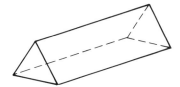

Look at the prisms shown here and count their faces, corners, edges. Put the results on your
table.

square prism

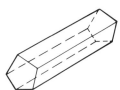
pentagonal prism

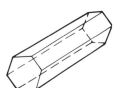
hexagonal prism

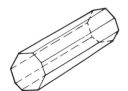
octagonal prism

	name	faces	corners	edges	faces+corners
	pyramid on a square base	5	5	8	10
1	pyramid on a pentagonal base				
2	pyramid on a hexagonal base				
3	pyramid on a heptagonal base				
	triangular prism	5	6	9	11
4	square prism				
5	pentagonal prism				
6	hexagonal prism				
7	octagonal prism				

LONDON'S UNDERGROUND

Central area (Simplified)

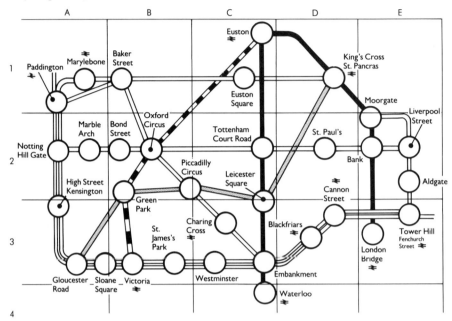

A This map shows 2 stations between Euston and Bank.

How many stations are there between

1 Waterloo and Tottenham Court Road?
2 Bank and Bond St?
3 Liverpool St and Marble Arch?
4 Embankment and Baker St?
5 Euston and Waterloo?

6 Cannon St and Gloucester Road?
7 Green Park and King's Cross?
8 Notting Hill Gate and Liverpool St?
9 Leicester Square and Sloane Square?
10 Moorgate and Marylebone?

B To go from Euston to Bond Street change at Oxford Circus.

Where do you change to go from

1 Euston to St Paul's?
2 King's Cross to Piccadilly Circus?
3 St Paul's to Leicester Square?
4 Oxford Circus to Victoria?
5 Baker St to Leicester Square?

6 Blackfriars to Piccadilly Circus?
7 King's Cross to Tottenham Ct Rd?
8 Liverpool St to Leicester Square?
9 Piccadilly Circus to Victoria?
10 Marble Arch to Waterloo?

C Oxford Circus is in the square 2B.
1 Which station is in 2C?
2 Which station is in 4C?
3 Which stations are in 2A?
4 Which stations are in 3E?
5 Which station is in 1B?

6 Where is Leicester Square?
7 Where is Oxford Circus?
Bank is on two different lines.
8 Which station is on most lines?
9 How many lines pass through Bond St?

94

NETWORKS

1

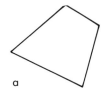 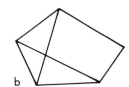 

a   b   c   d   e

Start at any corner and try to draw these shapes using the rules:

The pencil must not leave the paper.
The pencil must not go over a line twice.

It is possible to draw only 4 of them with these rules.

2

Here is another diagram which you cannot draw using the same rules.

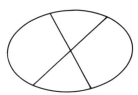

Look at each of the 5 joints.

One of them has 4 lines meeting.
Call it a 4-joint.

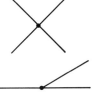

What would this joint be called?

4 of them have 3 lines meeting.
Call this kind a 3-joint.

3 Look at the shapes a to e, count their joints and complete the table.

Shapes which can be drawn

	2-joint	3-joint	4-joint	5-joint
a	4			
c				
d		2		
e	4	1	1	

Shapes which cannot be drawn

	2-joint	3-joint	4-joint	5-joint
b	1	4	1	

4 Now draw the following shapes, count their joints and add these results to the table. Look carefully at the 3-joint columns. This should help you to decide which shapes can be drawn and which cannot.

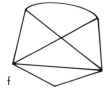    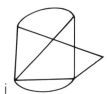

f   g   h   i   j

SUNSHINE

	Majorca		Ibiza		Alicante		Malaga		San Remo		Naples		Tangier	
April	19	8.0	20	8.4	21	8.6	21	7.9	18	6.7	19	5.9	18	7.8
May	24	9.0	23	9.5	24	9.7	27	9.7	20	8.6	23	8.2	22	10.2
June	27	10.7	27	11.2	27	9.6	27	10.8	24	9.4	27	9.3	24	10.9
July	29	11.6	29	12.2	30	10.6	29	11.4	27	11.2	30	11.3	27	11.8
Aug.	30	10.5	29	11.0	31	10.2	29	10.8	27	10.1	30	10.6	28	11.2
Sept.	27	7.9	28	8.3	28	8.0	27	8.3	25	8.0	27	6.5	26	8.1

In holiday towns the number of hours sunshine and the highest temperature for each day is recorded.

A holiday brochure showed the average temperature in degrees Celsius for each month.
Check that the table shows:
Naples has an average of 9.3 hours of sunshine each day in June and the average highest temperature is 27 °C.

1 Which is the hottest month in Majorca?

2 In which month did Majorca have most sunshine?

3 Which place has the best sunshine record for April?

4 Which place has the worst sunshine record for April?

5 Which place has the best sunshine record for September?

6 Where are you likely to get the warmest weather in September?

7 Which is the hottest place of all and when?

8 Which is the sunniest place of all and when?
   Are these the results you would expect?

9 In which month is there the biggest rise in temperature
   compared with the previous month? Which resort was it?

10 Some school children in Naples measured the hours of sunshine each day for five days. They
   got 8.0 hours, 8.1 hours, 8.5 hours, 8.4 hours, 8.0 hours.
   What was the average for that week?
   Which month do you think it was?

GROWING HEXAGONS

1 The smallest hexagon A has sides
which are 1 unit long. It encloses
six triangles.
The next hexagon B has sides which
are 2 units long.
How many triangles does it
enclose?

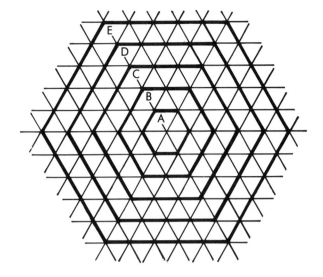

2 The hexagon C has sides 3 units long.
How many triangles does C enclose?

3 Copy and complete this table:

Length of side	Ratio of lengths	Number of triangles	Ratio of number triangles
A = 1		6	
B = 2	A:B = 1:2	24	A:B = 1:4
C = 3	A:C =		A:C =
D = 4	A:D =		A:D =
E = 5	A:E =		A:E =

Use the pattern for how the numbers grow to fill in more of the table:

F = 6	A:F = 1:6		A:F =
G = 7	A:G =		A:G =
H = 8	A:H =		A:H =

THE LAWN

A

1 What area would the bumper size treat?

2 Mr Smith wants to treat 35 sq.m of lawn.
How much will it cost him if he chooses the 5 kg size bags?
How much more would it cost to use the small bags?

3 Mr Black needs enough for 70 sq.m.
What will be the cheapest way of buying it?

4 Mr Jones has made this plan of his two lawns:

He says one of them is about 5 m wide and 10 m long, the other is 10 m long and 6 m wide.
What is their total area?
He reckons that he needs 15 kg of fertilizer and chooses the bumper size.
How much will be left if he buys the largest size?
How much more would it have cost him to buy the 5 kg bags?

SOAP AND SHAMPOO

B An industrial survey showed the quantities of toilet soap and shampoo manufactured in various countries. A unit called a metric tonne is used.
1 In which of these countries was most toilet soap produced?
2 In which country was most shampoo produced?

	Toilet soap	Shampoo
Belgium	3812	9474
Denmark	4474	6587
Finland	8253	4328
Greece	9000	3000
Norway	4395	1817
Portugal	4944	4210
Sweden	6170	6000

The countries surveyed are listed alphabetically in the table.
3 Make a list of the countries in order of toilet soap production, largest first.
4 Make another list showing the production of shampoo, largest first.

98

A NEW CAR

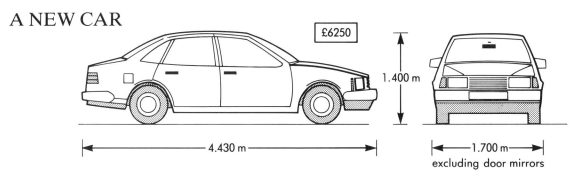

£6250

1.400 m

4.430 m

1.700 m
excluding door mirrors

Overall width including door mirrors is 1.876 m
Engine: 1389 cc
Maximum Speed: 109 mph
Fuel capacity: 61 litres

Petrol consumption (approximate)	Town driving	9.2 litres/100 km
	Constant speed 56 mph	5 litres/100 km
	Constant speed 75 mph	6.3 litres/100 km

1 Mr Jones is considering buying this car. His garage is 6 m long and the door
 opening is only 210 cm wide. Is it big enough for this car?
2 The hatchback model is 4.35 m long. How much longer or shorter is it than the 4-door model?
3 Why is this model called the 1.4 (1000 cc = 1 litre)?

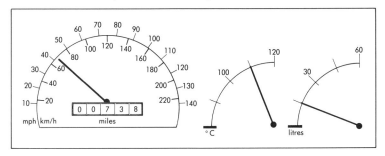

4 The diagram shows some of the instruments. How many miles has this car travelled?
5 What speed is 30 mph in km/h?
6 What is the maximum speed in km/h shown on the speedometer?
7 What will the mileometer show when the car has travelled one thousand five hundred miles?
8 How much petrol will the tank hold?
9 About how much petrol is the gauge showing?
10 How much more petrol could be put in the tank?
11 What is the fastest speed of this car in km/h?
12 Which kind of driving uses most petrol?
13 Most of Mr Jones's driving is done in the same town. He usually drives about 500 km each
 week. About how much petrol would this car need?
14 Which speed is most economical for petrol for a motorway journey?
 Mr Jones has to do a special trip on a motorway. He plans to travel 300 km.
15 What is the difference in the petrol he uses if he travels at about 50 mph or about 70 mph?
16 How much time would he save by travelling at 70 mph instead of 50 mph?
 Mr Jones decides to buy this car on HP.
 The HP terms are 20% deposit and £281.03 per month for 24 months.
17 How much is the deposit?
18 How much will he pay altogether for this car?

1

Scale: 5 cm represent 1 km

Coldtown
• C

• D
Downtown

The diagram shows a detail from a large map.
On this map CD = 45 cm.
  (i) Calculate the distance between
      C, Coldtown and D, Downtown.
 (ii) How much detail of the region will a
      map on this scale show?

2 (i) Use a calculator to find a: $a^3 + 12^3 = 1729$.
   (ii) There is another pair of whole numbers p and q: $p^3 + q^3 = 1729$
        Could either of them be more than 12?
        What is the largest value possible for one of them?
  (iii) Find the numbers p and q: $p^3 + q^3 = 1729$ (Making a list of the cubes of some whole
        numbers may be helpful)

3 Richard and Tom need to find the height of a tree. They use different methods:

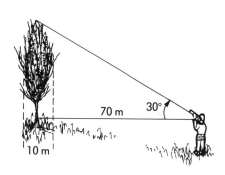

70 m   30°

10 m

Richard measures the distance on the ground below the tree and draws a diagram of it. He then
takes this photograph and measures the height and width of the tree on the photograph.
Use Richard's measurements to estimate the height of the tree.
Tom uses an instrument to measure the angle of elevation of the top of the tree.
Find the height of the tree from Tom's measurements (tan 30° = 0.58).

4 It is known that a certain rod is made of a metal which expands 4% of its length when heated to
a certain temperature. Gavin, Harry and John each measure the rod before and after it is
heated.

Correct the result which seems wrong.

Length	Cold	Hot
Gavin	50 cm	52 cm
Harry	500 mm	502 mm
John	500 mm	520 mm

5

In an experiment a ball was dropped into a thick liquid.
Its movement was timed as it fell:

Time, t sec	0	0.5	1.0	1.5	2.0	2.5
Depth	0		4	9		25

Which formula fits the results: $3t^2$, $4t^2$, $10t^2$?
Calculate the missing values.

100

1

THREAT TO DALE FOREST!
Plans to cut down 36% of the trees have just been announced.

There are 3200 trees in Dale Forest.
If the plans go ahead, how many trees will be left?

2 At a recent election 74% of the 50 500 electors voted. How many people voted?
The successful candidate polled 54% of the votes cast. How many votes was that?

3 When Mike went to Spain the rate of exchange was 205 pesetas for £1.
He exchanged £25. Should he get about 5000, 1000 or 500 pesetas?
He paid 4000 pesetas for a pair of sandals. Was that more or less than £20?

4 Mr Brown's clock loses 5 seconds every hour. How much will it lose in 24 hours?
It is set correctly on Monday at 8.00 pm.
What time will it be showing on Thursday at 8.00 pm?

5 What weight in pounds and ounces is shown on these dials? (16 oz = 1 lb)
    i)   ii)   iii)   iv)

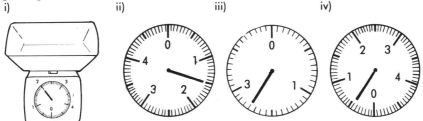

Which weights would be needed to make the same weighings on these scales?

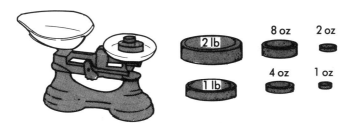

6 Find the area.

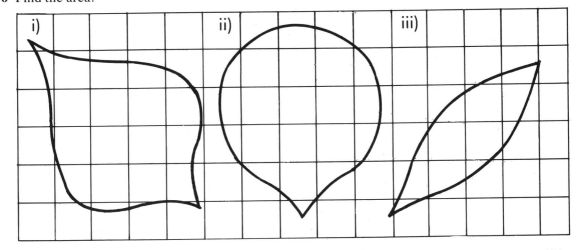

101

1 It is estimated that the average Briton uses 120 litres of water per day.

12% Washing machine	35% WC flushing	18% Baths, showers	35% Drinking, cooking, washing up

(i)   How much water is used daily for flushing the WC?
(ii) How many litres does the average Briton use for a bath or shower in a week?
(iii) How much water is used weekly in the washing machine?
(iv) If there are two people in the household, how much would be used weekly for drinking, cooking and washing up?

2 Mr Gray's electricity account showed he had used 1703 units in a quarter.
He thought it was high so he read his own meter at the beginning and end of a week:

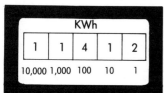 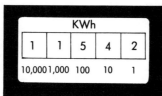

How many units had he used in that week?
Do these readings make it likely that he had been charged for too many units? (There are 13 weeks in a quarter)

3 In Italy Mr Johnson had to take a taxi from the airport to his hotel. It cost him 24 000 lire. The exchange rate was 2356 lire to the £1, so did he pay more or less than £10?
At the end of his stay Mr Johnson's hotel bill was 360 000 lire. About how much was that?

4 Some water has been boiled. As it cools down its temperature is taken every 5 minutes.
Copy this table showing the results. Finish the calculation of the fall in temperature.

Time	0 min	5 min	10 min	15 min	20 min	25 min	30 min	35 min
Temperature	100 °C	84 °C	71 °C	59 °C	50 °C	42 °C	35 °C	29 °C
Fall		16 °C	13 °C					

Which of these is likely to be the graph of these results? Why?

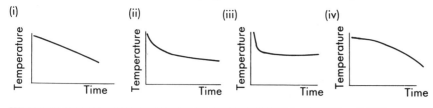

(i)   (ii)   (iii)   (iv)

5 This diagram shows the fuel used for domestic heating in Stoney Village. There are 420 houses and cottages.
How many are heated by coal?
How many are heated by gas?

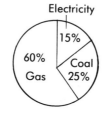

6 90 eggs were weighed with the following results:

Weight	52 g	53 g	54 g	55 g	56 g	57 g	58 g	59 g	60 g
Number of eggs	3	4	11	21	16	14	12	7	2

What is the total weight of all the eggs?
Draw the histogram of this distribution.
Calculate the mean weight of an egg.
How many eggs weighed less than the mean weight?

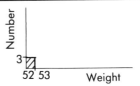

A A motorist has worked out a formula for the cost of running his car for a week.
It is £$(23 + 0.03n)$ where n is the number in kilometres he travels.
What is the cost when he travels 1950 km?

B There is a mistake in this table of values. Which value of x corresponds to the mistake, and what is the correct value?

x	0	0.5	1.0	1.5	2.0	2.5	3.0	3.5
$x^2 + 3x$	0	1.75	4.0	6.75	10	12.75	18.0	22.75

C A ship cruises between two ports, Windy Bay and Sunny Beach.
Sometimes it travels slowly for a holiday excursion. Other journeys are made faster and timed to allow passengers to catch a train.
The owner has noticed that the total cost to himself varies according to the average speed of the journey. The results are displayed in this graph:

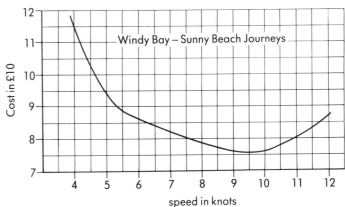

1 What is the slowest speed costed?
2 What is the fastest speed costed?
3 How much will it cost to travel at 7 knots?
4 How much will it cost to travel at 10.5 knots?
5 What is the most economical speed and how much does it cost?
6 If the cost has to be less than £80, what speeds are possible?
7 The owner needs to make a profit of 25% of his costs. A pleasure trip will not sail until there are 40 passengers paying full fare. If the pleasure trip travels at 5 knots, how much will it cost the owner?
How much must he charge each passenger to be sure of making the profit he wants?

1 In these models each brick above the first layer is placed exactly on another brick. What is the least possible number of bricks in each of these?

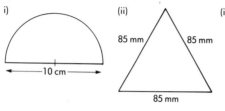 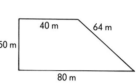

2 Calculate the perimeter of these shapes:
(Circumference of a circle = πd, use $\pi = 3.14$)

i)

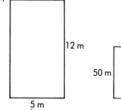

10 cm

(ii)

85 mm   85 mm

85 mm

(iii)

12 m

5 m

50 m

40 m   64 m

80 m

3 (i) Express as fractions: 90 out of 360,   60 out of 360,   120 out of 360.
  (ii) An arc is part of the circumference of a circle.
     Copy and complete these tables for circles of the circumferences shown:

(a) 180°

	Fraction	Circumference	Arc
		56 cm	
		64 cm	
		37 cm	

(c) 90°

	Fraction	Circumference	Arc
		10 cm	
		24 cm	
		30 cm	

(b) 60°

	Fraction	Circumference	Arc
		30 cm	
		54 cm	
		24 cm	

(d) 120°

	Fraction	Circumference	Arc
		18 cm	
		90 cm	
		66 cm	

4 A square sheet of metal is to be cut as shown in the diagram to make an open box with a square base. All the measurements are in mm.

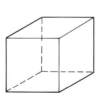 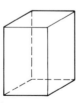 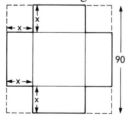 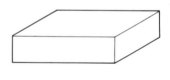

  (i) Make a sketch of the metal sheet to show the lengths if $x = 10$.
     Make a sketch of the box which would be formed. What is its volume?
  (ii) Sketch the box which would be formed if $x = 15$ mm. What is its volume?
  (iii) Sketch the box and find its volume if $x = 30$.
  (iv) What is the largest possible value of x?

This graph shows how the volume of the box changes as x increases.
  (v) For what values of x is the volume increasing?
  (vi) As x increases beyond 15 mm what happens to the volume?
  (vii) What is the maximum volume possible?

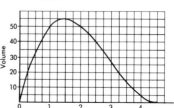

15 mixed practice papers

A
1. Write 45.36 correct to 2 significant figures.
2. Write 375.9 correct to the nearest ten.
3. What is the next whole number after 5609?
4. Write the next number in the sequence
   9, 18, 27, . . .
5. Write down the value of 5 in 2,514.
6. What is the next prime number after 13?
7. Find the product of 200 and 30.
8. Find the sum of 23.5 and 4.5.
9. Decrease £5 by 10%.
10. Make a true statement about 1.013 and 1.09.
11. $4.12 \div 4$.
12. 5^3.
13. $\sqrt{16}$.
14. 0.9×0.02.
15. $1.08 \div 0.2$.
16. $\frac{3}{8}$ as a decimal.
17. 29% as a decimal.
18. 5% of 30 m.
19. $2\frac{1}{2} \times 3$.
20. $70.5 - 16.81$.
21. A caterer cuts cakes into 12 equal portions. He needs 50 portions. How many whole cakes must he provide?
22. Write down the whole number values of x which make this true: $-2 < x < 3$.
23. Write down the even multiples of 5 below 30.
24. $A \times 5 = 2014$ and A is a whole number. How do you know this is the wrong answer?
25. Add five thousand and twenty one to fifteen thousand and seventy.

B
1. Write 36.25 correct to 3 significant figures.
2. Write 4526.7 correct to the nearest ten.
3. What is the next whole number after 7049?
4. Write the next number in the sequence
   8, 16, 24, . . .
5. Write down the value of 8 in 8376.
6. What is the next prime number after 19?
7. Find the product of 500 and 40.
8. Find the sum of 37.6 and 2.4.
9. Decrease £6 by 10%.
10. Make a true statement about 5.039 and 5.09.
11. $6.15 \div 3$.
12. 6^3
13. $\sqrt{36}$.
14. 0.3×0.03.
15. $2.08 \div 0.4$.
16. $\frac{5}{8}$ as a decimal.
17. 37% as a decimal.
18. 5% of 20 m.
19. $1\frac{1}{4} \times 5$.
20. $50.5 - 23.7$.
21. A caterer cuts cakes into 9 equal portions. He needs 50 portions. How many whole cakes must he provide?
22. Write down the whole number values of x which make this true: $-3 < x < 2$.
23. Write down the even multiples of 7 below 30.
24. $A \times 5 = 2010$ and A is an odd number. How do you know this is the wrong answer?
25. Add fourteen thousand and thirty five to four thousand and eighty.

105

A 1 Write down the largest factor of 15 other than 15.

2 What is fourteen thousand and eighty seven to the nearest hundred?

3 Arrange in order of size, longest first: 3.95 m, 400 cm, 3.9 m, 5000 mm.

4 Find the perimeter of a rectangle with sides 4.5 cm and 2 cm.

5 At noon the temperature was 4 °C. By 8.00 pm it had fallen 7 °C. What was the temperature at 8.00 pm?

6 The train leaves at 15.35. What is this in words?

7 Decrease £36 by 25%.

8 When the exchange rate is 203 pesetas to the pound, is 590 pesetas about £5, £3 or £6?

9 £50 prize money is divided in the ratio 2 : 3. How much is the larger prize?

10 A plan is drawn using 2 cm to represent 5 m. What does 7 cm represent?

11 A poster says 17 000 people are killed every year. How many is this per week?

12 Write the time 10 minutes before quarter to eleven.

13 An insect crawls 24 cm in 5 seconds. What speed is this?

14 The area of a square is 100 cm². How long is each side?

15 Use the formula $C = \pi d$ to find the circumference of a circle of diameter 10 cm ($\pi = 3.1$).

16 If $a = 3.5$ and $b = 4.2$, use a calculator to find the value of $(a + b) \times (b - a)$.

17 Use a calculator to express $5\frac{3}{16}$ as a decimal.

18 Two of the angles of a triangle are 55° and 80°. What is the size of the third angle?

19 A sequence of numbers is given by $3x$: 1.5, 4.5, 13.5. What is the next number?

20 If $x = 5$ and $y = 2$, find the value of $3x - y$.

B 1 Write down the smallest factor of 18 other than 1.

2 What is thirty-two thousand and twenty nine to the nearest hundred?

3 Arrange in order of size, longest first: 7.09 m, 6.97 m, 6000 mm, 700 cm.

4 Find the perimeter of a rectangle with sides 7.4 cm and 3 cm.

5 At noon the temperature was 12 °C. By 1.00 am it had fallen 8 °C. What was the temperature at 1.00 am?

6 The train leaves at 16.15. What is this in words?

7 Decrease £40 by 30%.

8 If the exchange rate is 198 pesetas to the pound, is 1000 pesetas about £5, £10 or £20?

9 £70 prize money is divided in the ratio 3 : 4. How much is the larger prize?

10 A plan is drawn using 2 cm to represent 5 m. What does 3 cm represent?

11 A poster says 17 000 people are killed every year. How many is this per month?

12 Write the time 10 minutes before quarter to one.

13 An insect crawls 15 cm in 2 seconds. What speed is this?

14 The area of a square is 36 cm². How long is each side?

15 Use the formula $C = \pi d$ to find the circumference of a circle of diameter 8 cm ($\pi = 3.1$).

16 If $a = 7.3$ and $b = 9.1$, use a calculator to find the value of $(a + b) \times (b - a)$.

17 Use a calculator to express $12\frac{9}{16}$ as a decimal.

18 Two of the angles of a triangle are 37° and 64°. What is the size of the third angle?

19 A sequence of numbers is given by $3x$: 0.7, 2.1, 6.3, ... What is the next number?

20 If $x = 7$ and $y = 4$, find the value of $2x - y$.

A
1 $2.79\,\text{m} \times 50$
2 $5x + 1 = 11$, find x
3 Simplify $3(4x - 1) + 2$
4 $3\frac{1}{4} \times 5$
5 0.3×0.02
6 Simplify $3a + 4 - a$
7 $3x^2 = 48$, find x
8 Express $450\,\text{g}$ in kg
9 $3\frac{7}{8}$ as a decimal
10 $4\,\text{m} \div 20$
11 Find the distance travelled in 3 minutes at $4\,\text{cm/sec}$.

12 On a map 1 cm represents 2 km. What does 5 cm represent?
13 Find 25% of 10 cm.
14 Find the perimeter of a square of side 6 cm.
15 What percentage is 4 km out of 50 km?
16 Find the area of a circle of radius 3 cm $(A = \pi r^2, \pi = 3.1)$.
17 Find the time taken to travel 30 m at 5 m/min.
18 If $a = 1.2$ and $b = 4$, find the value of $5a - b$.
19 Write the next number in the sequence: $1, 4, 9, 25, \ldots$
20 Write 15.25 hours using am or pm.

21

$a = 70°$, find x

22

$a = 123°$, find x

23

Find x

24
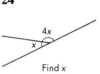
Find x

25

Find x

B
1 $5.31\,\text{m} \times 20$
2 $4x + 2 = 22$, find x
3 Simplify $5(2x - 1) + 3$
4 $2\frac{1}{4} \times 6$
5 0.4×0.03
6 Simplify $5x + 2 - x$
7 $2x^2 = 50$, find x
8 Express $560\,\text{g}$ in kg
9 $2\frac{3}{8}$ as a decimal
10 $6\,\text{m} \div 20$
11 Find the distance travelled in 2 minutes at $5\,\text{cm/sec}$.

12 On a map 1 cm represents 4 km. What does 6 cm represent?
13 Find 75% of 20 cm.
14 Find the perimeter of a square of side 5.5 cm.
15 What percentage is 2 km out of 20 km?
16 Find the area of a circle of radius 2 cm $(A = \pi r^2, \pi = 3.1)$.
17 Find the time taken to travel 500 m at 10 m/min.
18 If $a = 3.6$ and $b = 5$, find the value of $3a - b$.
19 Write the next number in the sequence: $100, 81, 64, \ldots$
20 Write 19.20 hours using am or pm.

21

$a = 105°$, find x

22

$a = 87°$, find x

23

Find x

24
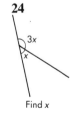
Find x

25
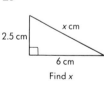
Find x

A
1 £2.79 + £4.76
2 0.7×0.8
3 Simplify $6(4 + 4a)$
4 $\frac{1}{3} - \frac{1}{4}$
5 0.3^2
6 $5x^2 - x^2$
7 $3x + 11 = 32$, find x
8 Express 450 mm in cm
9 If $x = 3$, find x^3
10 1.04×500
11 Find the mean of 6, 11 and 16.

12 Divide £35 in the ratio 2 : 5.
13 Express 50% as a fraction.
14 Find the perimeter of an equilateral triangle of side 7 cm.
15 What fraction is 6 of 18?
16 Find the area of a triangle of base 4 cm and height 3 cm.
17 Find the volume of a cuboid with sides 7 cm, 6 cm and 3 cm.
18 If $a = 7$ and $b = 5$, find the value of $2a - b$.
19 Which of these are multiples of 9: 99, 29, 18?
20 What fraction is 5 of 25?

21

$a=55°$, find x

22

$p=60°$, $q=70°$, find x

23

$t=37°$, find x

24

$k=80°$, find x

25

$a=85°$, $b=80°$, find x

B
1 £3.86 + £5.27
2 0.5×0.7
3 Simplify $5(3 + 3d)$
4 $\frac{1}{4} - \frac{1}{5}$
5 0.4^2
6 $3y^2 - y^2$
7 $9 + 5x = 44$, find x
8 Express 675 mm in cm
9 If $x = 5$, find x^3
10 1.6×500
11 Find the mean of 9, 13 and 18.
12 Divide £36 in the ratio 4:5.

13 Express 30% as a fraction.
14 Find the perimeter of an equilateral triangle of side 6 cm.
15 What fraction is 4 of 16?
16 Find the area of a triangle of base 5 cm and height 3 cm.
17 Find the volume of a cuboid with sides 7 cm, 5 cm and 2 cm.
18 If $a = 9$ and $b = 3$, find the value of $2a - b$.
19 Which of these are multiples of 7: 70, 14, 17?
20 What fraction is 5 of 15?

21

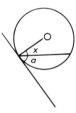

$a=64°$, find x

22

$a=40°$, $b=55°$, find x

23

$y=110°$, find x

24

$a=30°$, find x

25

$a=160°$, $b=70°$, $c=70°$, find x